T0281529

Mining Complex
Networks

Mining Complex Networks

Bogumił Kamiński

Paweł Prałat

François Théberge

CRC Press
Taylor & Francis Group
Boca Raton London New York

CRC Press is an imprint of the
Taylor & Francis Group, an **informa** business
A CHAPMAN & HALL BOOK

First edition published 2022
by CRC Press
6000 Broken Sound Parkway NW, Suite 300, Boca Raton, FL 33487-2742

and by CRC Press
2 Park Square, Milton Park, Abingdon, Oxon, OX14 4RN

CRC Press is an imprint of Taylor & Francis Group, LLC

ISBN: 978-1-032-11203-9 (hbk)
ISBN: 978-1-032-11205-3 (pbk)
ISBN: 978-1-003-21886-9 (ebk)

DOI: 10.1201/9781003218869

Publisher's note: This book has been prepared from camera-ready copy provided by the authors

Contents

Preface

Introduction

Data science is a multi-disciplinary field that uses scientific and computational tools to extract valuable knowledge from, typically, large data sets. Once the data is processed and cleaned, it is analyzed and presented in a form that is appropriate to support decision making processes. As collecting data has become much easier and cheaper these days than in the past, data science and machine learning tools have become widely used in companies of all sizes. Indeed, data-driven businesses were worth $1.2 trillion collectively in 2020, an increase from $333 billion in the year 2015, and it seems that this trend is going to persist in the future.

This book concentrates on mining networks, a subfield within data science. Virtually every human-technology interaction, or sensor network, generates observations that are in some relation with each other. As a result, many data science problems can be viewed as a study of some properties of complex networks in which nodes represent the entities that are being studied and edges represent relations between these entities. In these networks (for example, the Instagram on-line social network, the 4th most downloaded mobile app of the 2010s), nodes not only contain some useful information (such as the user's profile, photos, tags) but are also internally connected to other nodes (relations based on follower requests, similar users' behaviour, age, geographic location). Such networks are often large-scale, decentralized, and evolve dynamically over time. Mining complex networks in order to understand the principles governing the organization and the behaviour of such networks is crucial for a broad range of fields of study, including information and social sciences, economics, biology, and neuroscience. Here are a few selected typical applications of mining networks:

1. community detection (which users on some social media platform are close friends),

2. link prediction (who is likely to connect to whom on such platforms),

3. predicting node attributes (what advertisement should be shown to a given user of a particular platform to match their interests),

4. detecting influential nodes (which users on a particular platform would be the best ambassadors of a specific product).

After reading this book, one should be able to answer such questions, and much more, using state-of-the-art methods and computational techniques.

Target Audience

The book was written based on the lecture notes for a graduate course entitled *Graph Mining (DS 8014)* which was offered to students enrolled in the *Data Science and Analytics* Master's program at *Ryerson University* (Toronto, Canada). This textbook is aimed to be suitable for an upper-year undergraduate course or a graduate course. Students in programs such as data science, mathematics, computer science, business, engineering, physics, statistics, and social science will benefit from courses that are based on this textbook. Having said that, this book can be successfully used by all enthusiasts of data science at various levels of sophistication who would like to expand their knowledge or consider changing their career path. The Core Material (Part I) can be successfully used for a 12-week long course (for example, in Canadian system) but we additionally provide the Additional Material (Part II) that can be added for a 15-week long course (for example, in US or European systems).

Need for Another Book

This textbook is not the first (and certainly not the last) book related to network science. There are a number of excellent books, including those that we list in Section 1.15 that conceptually overlap with our book. Let us then present a few reasons why we decided to write this book.

Most books present a mixture of various topics in modelling and mining networks. Modelling complex networks is an important research direction and a few random graph models are included in our book but are mainly used as tools to benchmark and guide algorithms or to create synthetic networks for testing the behaviour of the tools in various scenarios. We focus on aspects related to mining complex networks, and carefully select the most important tools to create a nice and coherent blend that is appropriate for a one term course.

The three authors actively collaborate together, publishing research papers on various topics related to mining networks, including community detection algorithms, mining hypergraphs, unsupervised evaluation of graph embeddings, synthetic random graph models, anomaly detection algorithms, and link prediction algorithms. Our respective individual skills and experiences nicely complement each other, providing three different perspectives: pure mathematics (Paweł), mining large networks (Franois), and applying machine learning tools in business (Bogumił). This cumulative experience enables us to carefully select problems and tools that are suitable for a one-term course on mining networks. The content of this textbook represents the most important and useful aspects of the daily life of a data scientist, and with its use, data scientists can make a meaningful impact in business.

Most existing related books concentrate on theory. On the other hand, in our book the theoretical foundations are combined with practical experiments where students are expected to code and analyze graph datasets by themselves. This book is accompanied by Jupyter notebooks[1] (in Python and Julia) which not only contain all of the experiments presented in the book but which also include additional material. We will continue updating them, making sure they work with currently available environments. In particular, we use the `igraph`[2] library for Python which distinguishes us from other books that also use Python for their experiments. The `igraph` network analysis tool was chosen due to its superior performance in dealing with large graphs, and the richness of its library of graph analytics. For example, many centrality measures and graph clustering algorithms are available directly within `igraph`. Moreover, the library is written in C and can be used as such, and there are packages for R and Python, two of the most popular languages for data science. Moreover, we made publicly available videos that walk the reader through our notebooks which should be useful for readers that read the book by themselves and not as a part of a course offered at some university. Finally, we also made slides publicly available for the instructors to use, which should help them to adopt the textbook for their needs and their audience.

A distinguishing feature of mining networks, as opposed to traditional data mining, is that very often one needs to implement custom algorithms to perform an analysis for a given problem at hand. In traditional data mining, there are standard tools such as deep-learning networks, XGBoost, etc., to which we typically just pass appropriately prepared data. In mining networks, despite the fact that there exist standard tools and techniques, they usually require slight modifications to fit the studied problem. Because of this, apart from applying standard algorithms that are pre-implemented in the libraries such as `igraph`, one often needs to complement them with carefully tailored code that is computationally intensive. The reader will be able to notice this characteristic in virtually every chapter of this book. In such cases, one needs tools that allow one to implement such custom code efficiently while ensuring the code's speed (as usually complex networks are large). Traditionally, in such situations data scientists faced the so-called *two language problem*. In order to write the code efficiently Python was used, as it is a nice language for prototyping. However, these implementations were usually not scalable. Therefore,

---

[1]see `jupyter.org`; also available in Anaconda (`www.anaconda.com`) and other sources
[2]`igraph.org/python`

the next step was to re-write the prototype in some low level language such as C++.

In order to solve the two language problem, in this book we provide implementations of the examples not only using the Python language but also using the Julia language. Julia, like Python, is a high-level language (actually, in many cases the code is quite similar) but at the same time it is compiled (as opposed to Python which is interpreted), which allows the execution speed of the programs to be comparable to languages such as C++. These features of the Julia language have resulted in its popularity increasing recently, not only for mining complex networks but for all kinds of data science tasks that require performance and scalability.

About the Authors

Bogumił Kamiński is the Chairman of the Scientific Council for the Discipline of Economics and Finance at SGH Warsaw School of Economics. He is also an Adjunct Professor at the Data Science Laboratory at Ryerson University. Bogumił is an expert in applications of mathematical modelling to solving complex real-life problems. He is also a substantial open-source contributor to the development of the Julia language and its package ecosystem.

Paweł Prałat is a Professor of Mathematics at Ryerson University, whose main research interests are in random graph theory, especially in modelling and mining complex networks. He is the Director of Fields-CQAM Lab on Computational Methods in Industrial Mathematics at The Fields Institute for Research in Mathematical Sciences and has pursued collaborations with various industry partners as well as the Government of Canada. He has written more than 170 papers and 3 books with 130 plus collaborators.

François Théberge holds a B.Sc. degree in applied mathematics from the University of Ottawa, a M.Sc. in telecommunications from INRS, and a PhD. in electrical engineering from McGill University. He has been employed by the Government of Canada since 1996 during which he was involved in the creation of the data science team as well as the research group now known as the Tutte Institute for Mathematics and Computing. He also holds an adjunct professorial position in the Department of Mathematics and Statistics at the University of Ottawa. His current interests include relational-data mining and deep learning.

Accompanied Material

Additional complementary material can be found here

https://www.ryerson.ca/mining-complex-networks/

Part I

Core Material

1

Graph Theory

1.1 Notation

Before we move to graph theory, let us introduce some basic definitions and notation that will be used throughout this book. We will use $\mathbb{R}$ to denote the set of **real numbers**, $\mathbb{N} = \{1, 2, \ldots\}$ to denote the set of **natural numbers**, and $\mathbb{Z} = \{\ldots, -2, -1, 0, 1, 2, \ldots\}$ to denote the set of **integers**. For a given natural number ℓ, let $[\ell] := \{1, 2, \ldots, \ell\}$.

For a given $x \in \mathbb{R}$, the **floor function** $\lfloor x \rfloor$ is the largest integer less than or equal to x. Similarly, the **ceiling function** $\lceil x \rceil$ is the smallest integer that is greater than or equal to x. Finally, $\lfloor x \rceil$ rounds a real number x to its nearest integer, with the convention that for each $\ell \in \mathbb{Z}$, $\lfloor \ell + 1/2 \rceil = \ell + 1$.

For a given set S, the **power set** of S is the set of all subsets of S, including the empty set and S itself. The power set is usually denoted as $\mathcal{P}(S)$ or 2^S. It is easy to see that if S is finite, then $|\mathcal{P}(S)| = 2^{|S|}$. Similarly, for a given set S and $\ell \in [|S|] \cup \{0\}$, $\binom{S}{\ell}$ is the set consisting of all subsets of S of size ℓ, that is,

$$\binom{S}{\ell} = \left\{ T \subseteq S : |T| = \ell \right\}.$$

Clearly, if S is finite, then

$$\left| \binom{S}{\ell} \right| = \binom{|S|}{\ell},$$

and

$$\mathcal{P}(S) = \bigcup_{\ell=0}^{|S|} \binom{S}{\ell},$$

so

$$2^{|S|} = |\mathcal{P}(S)| = \left| \bigcup_{\ell=0}^{|S|} \binom{S}{\ell} \right| = \sum_{\ell=0}^{|S|} \left| \binom{S}{\ell} \right| = \sum_{\ell=0}^{|S|} \binom{|S|}{\ell},$$

something we already knew from the binomial theorem. Indeed, the binomial theorem says that for any $x, y \in \mathbb{R}$ and any $s \in \mathbb{N} \cup \{0\}$, we have

$$(x + y)^s = \sum_{\ell=0}^{s} \binom{s}{\ell} x^{s-\ell} y^{\ell}.$$

DOI: 10.1201/9781003218869-1

The above identity follows as a special case when $x = y = 1$.

Finally, let us mention about the **multinomial coefficient**, a useful generalization of the classical binomial coefficient that we used above. For any natural numbers $m_1, m_2, \ldots, m_n$ such that $\sum_{i=1}^{n} m_i = d \in \mathbb{N}$, we define it as follows:

$$\binom{d}{m_1, m_2, \ldots, m_n} = \frac{d!}{m_1!\, m_2! \cdots m_n!}.$$

The multinomial coefficients have a direct combinatorial interpretation. $\binom{d}{m_1, m_2, \ldots, m_n}$ is equal to the number of ways of depositing d distinct objects into n distinct bins, with m_1 objects in the first bin, m_2 objects in the second bin, and so on. It is indeed a generalization of the binomial coefficient as $\binom{d}{m} = \binom{d}{m, d-m}$.

1.2   Probability

We will also need some basic definitions and notions from probability theory. The set of possible outcomes of an experiment is called the **sample space** and is denoted by Ω. An **elementary event** is an event that contains only a single outcome in the sample space. For example, when a coin is tossed, there are two possible outcomes, "head" (represented by 1) and "tail" (represented by 0), and so $\Omega = \{0, 1\}$. If the coin is tossed n times, then there are 2^n possible outcomes that can be represented by the following sample space $\Omega = \{0, 1\}^n$, the set of binary vectors of length n.

Note that now we may think of **events** simply as subsets of Ω. For reasons beyond the scope of this book, if Ω is arbitrary, then its power set may be too large for probabilities to be assigned reasonably to all of its members. Fortunately, in this book, we typically deal with finite structures or instances when Ω is countable so this causes no problems. Hence, from now on we will assume that this assumption holds. In this scenario, a **probability measure** is a function $\mathbb{P} : \Omega \to [0, 1]$ such that $\sum_{\omega \in \Omega} \mathbb{P}(\omega) = 1$; $\mathbb{P}(\omega)$ is the probability that an elementary event ω holds. Then, the probability that an event $A \subseteq \Omega$ holds, is simply equal to $\mathbb{P}(A) = \sum_{\omega \in A} \mathbb{P}(\omega)$. The pair $(\Omega, \mathbb{P})$ is called a **probability space**. In our earlier example where the coin is tossed n times, each outcome occurs with the same probability, that is, $\mathbb{P}(\omega) = 2^{-n}$ for any $\omega \in \Omega$. Such probability spaces are called **uniform**.

If $\mathbb{P}(B) > 0$, then the **conditional probability** that A occurs given that B occurs is defined to be

$$\mathbb{P}(A|B) = \frac{\mathbb{P}(A \cap B)}{\mathbb{P}(B)}.$$

Events A, B are **independent** if

$$\mathbb{P}(A \cap B) = \mathbb{P}(A)\mathbb{P}(B).$$

In general, events $A_1, A_2, \ldots, A_n$ are independent if for any $I \subseteq [n]$,

$$\mathbb{P}\left(\bigcap_{i \in I} A_i\right) = \prod_{i \in I} \mathbb{P}(A_i).$$

Intuitively, the property of independence means that the knowledge of whether some of the events $A_1, A_2, \ldots, A_n$ occurred does not affect the probability that the remaining events occur.

A (real) **random variable** on a probability space $(\Omega, \mathbb{P})$ is a function $X : \Omega \to \mathbb{R}$. The **expectation** of a (real) random variable X is defined as follows:

$$\mathbb{E}[X] = \sum_{\omega \in \Omega} \mathbb{P}(\omega) X(\omega).$$

We say that two random variables X and Y are independent if, for all $x, y \in \mathbb{R}$, events $\{\omega \in \Omega : X(\omega) = x\}$ and $\{\omega \in \Omega : Y(\omega) = y\}$ are independent, that is:

$$\mathbb{P}(X = x \wedge Y = y) = \mathbb{P}(X = x) \cdot \mathbb{P}(Y = y). \tag{1.1}$$

This definition can be naturally extended to more than two random variables.

Coming back to our example, let X be the random variable counting the number of heads in n independent coin tossings. Let X_i denote the outcome of the i-th coin tossing: $X_i = 1$ ("head") with probability $p \in [0, 1]$; otherwise, $X_i = 0$ ("tail"). Such a random variable is called the **Bernoulli random variable**. Clearly $X = \sum_{i=1}^n X_i$. A random variable formed as the sum of independent Bernoulli random variables is called the **binomial random variable** with the probability of success (that is, getting a head) equal to p, and denoted by $\text{Bin}(n, p)$. Now, as X_i are independent, from (1.1) we get that:

$$\mathbb{P}(X = k) = \binom{n}{k} p^k (1 - p)^{n-k}, \qquad 0 \le k \le n,$$

and so

$$\begin{aligned} \mathbb{E}[X] &= \sum_{k=0}^n \mathbb{P}(X = k) \cdot k = 0 + \sum_{k=1}^n \binom{n}{k} p^k (1 - p)^{n-k} \cdot k \\ &= np \sum_{k=0}^{n-1} \binom{n-1}{k} p^k (1 - p)^{n-1-k} = np, \end{aligned}$$

by the binomial theorem. The expectation of $\text{Bin}(n, p)$ can be obtained slightly easier using the **linearity of expectation** that claims that for any two random variables X, Y and $a, b \in \mathbb{R}$,

$$\mathbb{E}[aX + bY] = a\,\mathbb{E}[X] + b\,\mathbb{E}[Y].$$

For our example ($X \in \mathrm{Bin}(n, p)$) we observe that $\mathbb{E}[X_i] = 1 \cdot p + 0 \cdot (1 - p) = p$ for all i, and so we get that

$$E[X] = E\left[\sum_{i=1}^{n} X_i\right] = \sum_{i=1}^{n} E[X_i] = \sum_{i=1}^{n} p = np.$$

The **variance** of random variable X is defined as follows:

$$\begin{aligned} \mathrm{Var}[X] &= \mathbb{E}\left[(X - \mathbb{E}[X])^2\right] = \mathbb{E}[X^2] - (\mathbb{E}[X])^2 \\ &= \sum_{\omega \in \Omega} (X(\omega) - \mathbb{E}[X])^2\, \mathbb{P}(\omega). \end{aligned}$$

The variance has the following properties: $\mathrm{Var}[X] \geq 0$,

$$\mathrm{Var}[aX + b] = a^2\, \mathrm{Var}[X],$$

and

$$\begin{aligned} \mathrm{Var}[X + Y] &= \mathbb{C}\mathrm{ov}[X, X] + 2\, \mathbb{C}\mathrm{ov}[X, Y] + \mathbb{C}\mathrm{ov}[Y, Y] \\ &= \mathrm{Var}[X] + 2\, \mathbb{C}\mathrm{ov}[X, Y] + \mathrm{Var}[Y], \end{aligned}$$

where the **covariance** $\mathbb{C}\mathrm{ov}[X, Y]$ is defined as follows:

$$\mathbb{C}\mathrm{ov}[X, Y] = \mathbb{E}\left[(X - \mathbb{E}[X])(Y - \mathbb{E}[Y])\right] = \mathbb{E}[XY] - \mathbb{E}[X]\,\mathbb{E}[Y].$$

Finally, assuming that both X and Y have positive variance, the **Pearson's correlation coefficient**, a statistic that measures the linear correlation between two random variables X and Y, is defined as follows:

$$\rho_{X,Y} = \frac{\mathbb{C}\mathrm{ov}[X, Y]}{\sqrt{\mathrm{Var}[X]\,\mathrm{Var}[Y]}}. \tag{1.2}$$

As a direct consequence of the **Cauchy–Schwarz inequality**, we get that $\rho_{X,Y}$ can only take values from the interval $[-1, 1]$. An extreme value of 1 implies that there exists a constant $\alpha > 0$ such that $\alpha(X - \mathbb{E}[X]) = Y - \mathbb{E}[Y]$. The other extreme value of -1 implies that the same relationship holds for some $\alpha < 0$. Note that, as X and Y have positive variance, it is not possible that $\alpha = 0$. A value of 0 implies that there is no linear correlation between the two variables. Note that these variables do not have to be independent but, if random variables X and Y are independent, then $\mathbb{C}\mathrm{ov}[X, Y] = \rho_{X,Y} = 0$.

Let us use these facts to calculate $\mathrm{Var}[X]$, where $X \in \mathrm{Bin}(n, p)$. As above, let X_i denote the independent Bernoulli random variable with probability of success equal to p. First, note that

$$\mathbb{V}\mathrm{ar}[X_i] = E[X_i^2] - (E[X_i])^2 = p - p^2 = p(1 - p).$$

Now, since $\mathbb{C}\mathrm{ov}[X_i, X_j] = 0$ for $i \neq j$ we get that

$$\mathrm{Var}[X] = \mathrm{Var}\left[\sum_{i=1}^{n} X_i\right] = \sum_{i=1}^{n} \mathrm{Var}[X_i] = np(1 - p).$$

1.3 Linear Algebra

In order to describe some of the models that we introduce in this book, we will use tools from linear algebra. In this section, we introduce some basic notation that we will use and recall a few key theorems that we will need. A reader interested in a more in-depth material is encouraged to look at one of the many textbooks on linear algebra.

We typically denote vectors using bold lowercase letters such as $\mathbf{x}$, and always consider them to be column vectors. Matrices are denoted using bold uppercase letters such as $\mathbf{A}$, and all matrices we use in this book have real valued elements. The **identity matrix** (of size n) is the $n \times n$ square matrix with ones on the main diagonal and zeros elsewhere, and is denoted as $\mathbf{I}_n$ or simply $\mathbf{I}$ when the size is determined by the context. When A is $m \times n$ matrix, then

$$\mathbf{I}_m\mathbf{A} = \mathbf{A}\mathbf{I}_n = \mathbf{A}.$$

The **transpose** of a matrix $\mathbf{A}$, denoted by $\mathbf{A}^T$, flips a matrix over its diagonal, that is, it switches the row and column indices of the matrix $\mathbf{A}$. Formally, $\mathbf{A}^T = (a_{ij}^T) = (a_{ji})$. A square matrix whose transpose is equal to itself is called symmetric; that is, $\mathbf{A}$ is **symmetric** if $\mathbf{A}^T = \mathbf{A}$. There are many interesting and important properties that are related to transposing matrices. For example, $(\mathbf{A}^T)^T = \mathbf{A}$, and $(\mathbf{A}\mathbf{B})^T = \mathbf{B}^T\mathbf{A}^T$.

The **determinant**, denoted $\det(\mathbf{A})$, is a real number that can be computed from the elements of a square matrix $\mathbf{A}$ and encodes certain properties of the linear transformation described by $\mathbf{A}$. Geometrically, it can be viewed as the signed volume scaling factor of the corresponding linear transformation. More importantly from our applications point of view, the determinant $\det(\mathbf{A})$ can be used to solve a system of linear equations represented by matrix $\mathbf{A}$, although other methods of solution are much more computationally efficient. The determinant has many properties such as $\det(\mathbf{I}) = 1$, $\det(\mathbf{A}^T) = \det(\mathbf{A})$, and $\det(\mathbf{A}\mathbf{B}) = \det(\mathbf{A}) \cdot \det(\mathbf{B})$.

Given a square matrix $\mathbf{A}$, if for some λ (possibly complex) there exists a vector $\mathbf{x}$ such that $\mathbf{A}\mathbf{x} = \lambda\mathbf{x}$, then we call λ an **eigenvalue** of a matrix and $\mathbf{x}$ its **eigenvector**. Equation $\mathbf{A}\mathbf{x} = \lambda\mathbf{x}$ can be equivalently stated as

$$(\mathbf{A} - \lambda\mathbf{I})x = \mathbf{0},$$

where $\mathbf{I}$ is the identity matrix and $\mathbf{0}$ is the zero vector. In turn, this equation has a non-zero solution $\mathbf{x}$ if and only if

$$\det(\mathbf{A} - \lambda\mathbf{I}) = 0. \tag{1.3}$$

It can be shown that the left hand of (1.3) is a polynomial function of the variable λ of degree n, the order of the matrix A. This polynomial is called the **characteristic polynomial** of $\mathbf{A}$. The **Fundamental Theorem of Algebra** implies that there are exactly n roots of the characteristic polynomial:

$\lambda_1, \lambda_2, \ldots, \lambda_n$. The roots may be real numbers but in general are complex numbers. Moreover, they may not all have distinct values.

It can be shown that $\det(\mathbf{A}) = \prod_{i=1}^{n} \lambda_i$. The **trace** $\operatorname{tr}(\mathbf{A})$ is, by definition, the sum of the diagonal entries of matrix $\mathbf{A}$ but it can be shown that $\operatorname{tr}(\mathbf{A}) = \sum_{i=1}^{n} \lambda_i$. Also note that eigenvalues of a matrix are equal to the eigenvalues of its transpose, since they share the same characteristic polynomial. Finally, it is known that if $\mathbf{A}$ is symmetric, then all its eigenvectors are **orthogonal** (that is, the inner product of any pair of them is zero) and all its eigenvalues are real numbers.

An $n \times n$ square matrix $\mathbf{A}$ is called **invertible** if there exists an $n \times n$ square matrix $\mathbf{B}$ such that

$$\mathbf{AB} = \mathbf{BA} = \mathbf{I}_n.$$

If this is the case, then the matrix $\mathbf{B}$ is uniquely determined by $\mathbf{A}$, and is called the **inverse** of $\mathbf{A}$, denoted by $\mathbf{A}^{-1}$. It can be shown that $\mathbf{A}$ is invertible if and only if $\det(\mathbf{A}) \neq 0$.

If $\mathbf{A}^{-1} = \mathbf{A}^T$ then we call a square matrix $\mathbf{A}$ **orthogonal**. As already mentioned above, if for two vectors $\mathbf{x}$ and $\mathbf{y}$ we have $\mathbf{x}^T \mathbf{y} = 0$ we call these vectors orthogonal. If additionally $\det(\mathbf{A}) = 1$, then we call $\mathbf{A}$ a **rotation** matrix, and if $\det(\mathbf{A}) = -1$, then $\mathbf{A}$ is called **roto-reflection** matrix (also called **improper rotations**). As the name suggests, a rotation matrix is a transformation matrix that is used to perform a rotation in the corresponding Euclidean space.

We call a square matrix $\mathbf{A}$ **irreducible** if, when it is considered as an adjacency matrix of a directed graph (where each non-zero entry represents an edge), it induces a strongly connected graph, that is, there exists a directed path between any two nodes of this graph. (Adjacency matrix is formally introduced in Section 1.5; see Section 1.7 for a formal definition of strongly connected graphs.) If an irreducible matrix additionally has only non-negative elements, then following the **PerronFrobenius theorem** the following properties hold: a) its eigenvalue that has the largest norm is real and positive—we call it the **leading eigenvalue**, b) there is exactly one eigenvector (up to multiplication) associated with this eigenvalue whose components are all positive and this is the only such eigenvector of this matrix.

Given any real matrix $\mathbf{A}$, we call $\mathbf{G} = \mathbf{A}^T \mathbf{A}$ the **Gram** matrix (or **Gramian** matrix). Notably, $\mathbf{G}$ is symmetric and has non-negative eigenvalues. As a consequence, $\mathbf{x}^T \mathbf{G} \mathbf{x} \geq 0$ for any vector $\mathbf{x}$. If we denote by $\lambda_{\min}$ and $\lambda_{\max}$ the minimum and, respectively, the maximum eigenvalue of $\mathbf{G}$, then it can be shown that for any non-zero vector $\mathbf{x}$ we have $\lambda_{\min} \leq \mathbf{x}^T \mathbf{G} \mathbf{x} / (\mathbf{x}^T \mathbf{x}) \leq \lambda_{\max}$.

1.4 Definition

A (**undirected**) **graph** $G = (V, E)$ consists of V, the **set of nodes**, and E, the **set of edges**. Each edge $uv \in E$ has two **endpoints**, u and v, and so each edge in G is formally a subset of V of size two. If $uv \in E$, then we say that u is **adjacent** to v. Note that under this definition the order of nodes when specifying an edge is not important: uv and vu refer to the same edge.

In particular, we will usually assume that the graph is **simple**, meaning there are no **loops** (edges from v to itself) nor **parallel edges** (multiple edges between v and u). Hence,

$$E \subseteq \binom{V}{2} = \left\{ e = uv = \{u, v\} : e \subseteq V, |e| = 2 \right\}.$$

We will occasionally deal with multi-graphs (graphs where loops and parallel edges are allowed), but such situations will be clearly marked.

A **directed graph** $D = (V, E)$ consists of the set of nodes and, respectively, the set of edges. However, this time each edge $uv \in E$ is a directed edge from u to v. Formally, each edge $uv \in E$ is then an ordered pair (u, v), and we say that an edge uv goes from node u to node v. Hence, the order of nodes when specifying an edge is important, as (u, v) is not the same as (v, u). As before, we assume that the graph is simple and so

$$E \subseteq \left\{ e = uv = (u, v) : u \in V, v \in V \setminus \{u\} \right\}.$$

1.5 Adjacency Matrix

Let $D = (V, E)$ be any directed graph on n nodes. Alternatively, let $G = (V, E)$ be any graph on n nodes. An **adjacency matrix** $\mathbf{A} = (a(u, v))_{u,v \in V}$ is a square matrix used to represent a graph. The elements of the matrix indicate whether pairs of nodes are adjacent or not. In the case of simple and unweighted graphs (regardless whether they are directed or undirected), we use $a(u, v) = 1$ to indicate that there is an edge from u to v; otherwise, $a(u, v) = 0$. As a result, the adjacency matrix is a $\{0, 1\}$-matrix with zeros on its diagonal. We will generalize this notion to weighted graphs in the next section. Clearly, if the graph is undirected, then the adjacency matrix is symmetric.

There are several other ways to represent a graph in a computer such as **edge list**, **adjacency list**, and **incidence matrix**. Some situations or algorithms that one wants to run on a graph call for one representation but others require a different representation. Moreover, the decision which representation should be used to achieve the best complexity of the algorithm usually depends on whether the graph that needs to be dealt with is sparse or dense. Fortunately, in high level languages such as Python or Julia, there are efficient libraries that allow one to work with graphs without deciding about the choice of the best data structure for a given algorithm. In the Jupyter notebooks associated with this book, we use the `igraph` library for Python and the `LightGraphs.jl` package for Julia. However, if more advanced programmers want to explicitly work with a given data structure, then both languages provide built-in modules and installed packages to do so. In particular, for many algorithms working on sparse graphs, it is convenient to store adjacency matrix in a sparse matrix. Sparse matrices are available in Python in the `SciPy` package and they are built-in into Julia standard installation. As a result, both Python and Julia can take advantage of the fact that almost all complex networks are sparse to improve the computation time of the performed experiments. As a special feature of the Julia, it is possible to pass an information to an algorithm about additional properties of the involved data structures such as the fact that the associated adjacency matrix is symmetric if the corresponding graph is undirected. The language can then take advantage of this fact and further improve the performance.

1.6 Weighted Graphs

Let $D = (V, E)$ be any directed graph on n nodes. Alternatively, let $G = (V, E)$ be any graph on n nodes. In weighted graphs, the value of $a(u, v)$ for an edge $uv \in E$ between nodes u and v (or from node u to node v, if the graph is directed) can take arbitrary positive real number; we usually avoid using weight zero as it is reserved to indicate that there is no edge between u and v. It is often mathematically convenient to assume that $0 < a(u, v) \leq 1$ but this is not a hard constraint and one can be flexible, if needed. As mentioned in the previous section, for unweighted directed graphs, we will assume that $a(u, v) = 1$ if there is an edge from u to v (or between u and v for undirected graphs).

The interpretation of the weights, however, is very important and it could vary a lot depending on the context or application in mind. A natural interpretation for weights is to represent some notion of distance (or dissimilarity) between the involved nodes. In that case, we can generalize notions such as a shortest path between nodes to take the weights into account. (Paths will be formally defined in the next section.) On the other hand, this has the opposite effect for measures of centrality: a node that is far from every other node is usually considered less central.

It is also possible for edge weights to measure some notion of similarity or closeness between the nodes; for example, if the weights correspond to the bandwidth in a communication network, then high values correspond to tightly connected nodes. Another example, as we will see in Section 3.7, is for weights to represent the amount of exchange between nodes such as the number of passengers. In those examples, higher edge weights often cause incident nodes to be considered more central. In such cases, the notion of path length is not clear anymore: we might consider the length of a path to be inversely proportional to the edge weights in some sense, or just reduce it to the unweighted definition of path length (the number of edges on a shortest path between two nodes).

1.7 Connected Components and Distances

Let $D = (V, E)$ be any directed graph on n nodes. Alternatively, let $G = (V, E)$ be any graph on n nodes. Let us also fix a pair of two nodes, $u, v \in V$. A sequence of nodes $P = (w_0, w_1, \ldots, w_\ell)$ is called a **path** from u to v if $w_0 = u$, $w_\ell = v$, and $w_{i-1}w_i \in E$ (that is, $a(w_{i-1}, w_i) > 0$) for all $i \in [\ell]$. The **length** of a path P is defined as the weight of the path, that is, it is equal to $\sum_{i \in [\ell]} a(w_{i-1}, w_i)$. In particular, for unweighted graphs it is simply equal to the number of edges P consists of.

A **connected component** of an undirected graph G is a maximal subgraph in which any two nodes are connected to each other by a path. If G has precisely one connected component, then we say that G is **connected**; otherwise, we say that G is **disconnected**. The same definition applies to a directed graph D, but in this case we say that it defines a **strongly connected component**, as there are some other natural generalizations to directed graphs.

The **distance** $\mathrm{dist}(u, v)$ from node u to node v is defined as the minimum length of a path (directed or undirected) from u to v, that is,

$$\mathrm{dist}(v, u) = \min_{(v=w_0, w_1, \ldots, w_\ell = u)} \sum_{i \in [\ell]} a(w_{i-1}, w_i), \qquad (1.4)$$

where the minimum is taken over all paths from v to u. In particular, $\mathrm{dist}(v, v) = 0$ as there is a degenerate path $P = (v)$ from v to v of length 0. Note that $\mathrm{dist}(u, v)$ is not defined when the two nodes belong to two different connected components. Having said that, there are many natural extensions to disconnected graphs such as setting $\mathrm{dist}(u, v) = n$ for such pairs of nodes. Note that the proposed weight of n is slightly larger than $n - 1$, the maximum possible distance between two nodes within the same connected component (for both weighted and unweighted graphs).

Finally, the **diameter** of a strongly connected directed graph D (or a connected undirected graph G) is defined as the maximum distance between two nodes, that is,

$$\mathrm{diam}(D) = \max_{u, v \in V} \mathrm{dist}(u, v).$$

Having said that, let us mention that the `igraph` library generalizes this definition to disconnected graphs. In this generalization, one simply takes the maximum distance $\mathrm{dist}(u, v)$ over all pairs of nodes u, v that belong to the same strongly connected component. Though this definition is neither natural nor is it always intuitive (in some situations, the diameter may decrease after removing some edges), we sporadically use it in our experiments later on.

1.8   Degree Distribution

Let $G = (V, E)$ be any graph on n nodes and m edges. For a given node $v \in V$, let $N(v)$ be the set of **neighbours** of v, that is,

$$N(v) = \{u \in V : uv \in E\} = \{u \in V : a(u, v) > 0\}.$$

Similarly, let $N_\ell(v)$ and $N_{\leq \ell}$ be the set of nodes at distance exactly ℓ and, respectively, at distance at most ℓ, that is,

$$
\begin{aligned}
N_\ell(v) &= \{u \in V : \mathrm{dist}(u, v) = \ell\} \\
N_{\leq \ell}(v) &= \{u \in V : \mathrm{dist}(u, v) \leq \ell\}.
\end{aligned}
$$

Let $\deg(v) = \sum_{u \in V} a(v, u)$ be the **degree** of a node v. For unweighted graphs, $\deg(v) = |N(v)|$ is simply equal to the number of neighbours of v; for weighted ones, it is a sum of weights of edges incident to v. The **degree sequence** is simply the sequence $\mathbf{d} = (\deg(v))_{v \in V}$. In order for the degree sequence to be well-defined, we always assume that the degree sequence is non-decreasing, that is,

$$\mathbf{d} = (\deg(v_1), \deg(v_2), \ldots, \deg(v_n))$$

with $\deg(v_1) \leq \deg(v_2) \leq \ldots \leq \deg(v_n)$.

The **degree distribution** d_ℓ of a graph is defined to be the fraction of nodes with degree ℓ, that is, $d_\ell = n_\ell / n$, where n_ℓ is the number of nodes of degree ℓ. It is often more convenient (especially for weighted graphs) to consider $d_{\leq \ell}$, the fraction of nodes with degree at most ℓ. It will be also convenient to use

$$\langle k \rangle = \frac{1}{n} \sum_{v \in V} \deg(v) = \sum_{\ell \in \mathbb{N}} \ell \, d_\ell = \frac{2m}{n}$$

to denote the **average degree** in the unweighted graph G. In general, for any $s \in \mathbb{N}$, the sth moment of the degree sequence $(\deg(v))_{v \in V}$ is defined as follows:

$$\langle k^s \rangle = \frac{1}{n} \sum_{v \in V} \deg(v)^s = \sum_{\ell \in \mathbb{N}} \ell^s \, d_\ell.$$

Using the language from probability theory, note that if X is a node taken uniformly at random from the set of nodes V, then we can simply write $\langle k^s \rangle = \mathbb{E}[\deg(X)^s]$.

Clearly, not every sequence of non-negative integers $\mathbf{d} = (d_1, d_2, \ldots, d_n)$ can occur as a degree sequence of some graphs. For example, since in any graph $G = (V, E)$ we have $\sum_{v \in V} \deg(v) = 2|E|$, it is a trivially necessary condition (but not sufficient one) that $\sum_{i \in [n]} d_i$ be even. We will come back to this issue and provide the necessary and sufficient condition in (2.5). In any case, this shows that it is important to recognize which sequences are feasible and which ones are not. A sequence of numbers is said to be a **graphic sequence** if one can construct a graph having the sequence as its degree sequence.

We will also need the following related definitions: for a given set of nodes $S \subseteq V$, the **volume of set** S is defined as $\mathrm{vol}(S) = \sum_{v \in S} \deg(v)$. In particular, the **volume of a graph** $G = (V, E)$ with m edges is equal to $\mathrm{vol}(V) = \sum_{v \in V} \deg(v) = 2m$. Finally, we define the **minimum** and the **maximum** degree of a graph G as $\delta(G) = \min_{v \in V} \deg(v)$ and, respectively, $\Delta(G) = \max_{v \in V} \deg(v)$.

Let $D = (V, E)$ be any directed graph on n nodes. One can easily generalize the notion of neighbourhoods to directed graphs but this time we have two types of neighbours: nodes that point towards v and nodes that v points to. As a result, we distinguish between in-neighbours and out-neighbours. In particular, we define the following two sets:

$$
\begin{aligned}
N^{in}(v) &= \{u \in V : uv \in E\} = \{u \in V : a(u, v) > 0\} \\
N^{out}(v) &= \{u \in V : vu \in E\} = \{u \in V : a(v, u) > 0\},
\end{aligned}
$$

and the two corresponding types of degrees: $\deg^{in}(v) = \sum_{u \in V} a(u, v)$ and $\deg^{out}(v) = \sum_{u \in V} a(v, u)$.

1.9 Subgraphs

A graph $G' = (V', E')$ is a **subgraph** of $G = (V, E)$ if $V' \subseteq V$ and $E' \subseteq E$. G' is a **spanning subgraph** if $V' = V$. If $V' \subseteq V$, then

$$
G[V'] = (V', \{uv \in E : u, v \in V'\})
$$

is the subgraph of G **induced** by V'. Similarly, if $E' \subseteq E$ then $G[E'] = (V', E')$ where

$$
V' = \{v \in V : \exists e \in E' \text{ such that } v \in e\}
$$

is also an **induced** subgraph of G (by E').

1.10   Special Families

Let $G = (V, E)$ be any graph. The **complement** of G is the graph $\overline{G} = (\overline{V}, \overline{E})$ with node set $\overline{V} = V$ and edge set $\overline{E}$ defined as follows: $uv \in \overline{E}$ if and only if $uv \notin E$.

Earlier in this chapter, we defined paths of length ℓ. Let us now introduce a few more special families of graphs which are often of interest, including paths (for completeness). Let $G = (V, E)$ be any graph and consider the following structures defined on subsets of its nodes:

- A sequence of distinct nodes $P_{\ell+1} = (w_0, w_1, \ldots, w_\ell)$ is called a **path** of length ℓ if $w_{i-1} w_i \in E$ for each $i \in [\ell]$.

- A sequence of distinct nodes $C_\ell = (w_1, w_2, \ldots, w_\ell)$ is called a **cycle** of length ℓ if $w_i w_{i+1} \in E$ for each $i \in [\ell - 1]$ and $w_\ell w_1 \in E$.

- A set of nodes $K_\ell = \{w_1, w_2, \ldots, w_\ell\}$ is called a **complete graph** on ℓ nodes (sometimes also called a **clique**) if it induces a set of pairwise adjacent nodes, that is, if $w_i w_j \in E$ for all $i, j \in [\ell]$ such that $i < j$.

- A set of nodes $\overline{K_\ell} = \{w_1, w_2, \ldots, w_\ell\}$ is called an **empty graph** on ℓ nodes (sometimes also called an **independent set**) if it induces a set of pairwise non-adjacent nodes, that is, if $w_i w_j \notin E$ for all $i, j \in [\ell]$ such that $i < j$.

A graph $G = (V, E)$ is **bipartite** if $V = X \cup Y$, where $X \cap Y = \emptyset$ and every edge $xy \in E$ satisfies $x \in X$ and $y \in Y$. Subsets X and Y of the set of nodes are called **partite sets**. A **complete bipartite graph** $K_{m,n} = (X \cup Y, E)$ is the graph with partite sets X, Y with $|X| = m$, $|Y| = n$, and edge set

$$E = \{xy : x \in X, y \in Y\}.$$

Clearly, $|E| = mn$.

A **forest** is a graph with no cycle. A **tree** is a connected forest (every connected component of a forest is a tree). A **leaf** is a node of degree 1. A **spanning tree** of a connected graph $G = (V, E)$ is a spanning subgraph $G' = (V, E')$ of G that is a tree (note that it implies that G' is connected). In particular, a path P_n and a **star** $K_{1,n-1}$ are examples of trees on n nodes. Each tree on n nodes has $n - 1$ edges.

Let $d \in \mathbb{N} \cup \{0\}$ and $n \in \mathbb{N}$ such that $n \geq d + 1$. We say that $G = (V, E)$ is a d-**regular graph** if $\deg(v) = d$ for each $v \in V$. Since the total volume of any graph is even, n has to be even if d is odd. If $n = d + 1$, then there is only one d-regular graph, namely, the complete graph K_{d+1}. On the other hand, if n is much larger than d, then there are many of them. We will consider this family of graphs, for example, when discussing random d-regular graphs in Section 2.6.

1.11 Clustering Coefficient

The clustering coefficient is a graph parameter that tries to measure the tendency for nodes to cluster together. It has been experimentally verified that many real-world networks (including social networks) consist of tightly connected groups of nodes; the likelihood of seeing such behaviour in a random network, in which each pair of nodes is independently connected with a relatively small probability, is very low. This natural phenomena can be informally described as "friends of my friends are often my friends too." The presence of such structure highly affects many important processes occurring in networks such as promotion of products via viral marketing or the spreading of computer viruses and infectious diseases. Formally, there are two versions of the clustering coefficient: the global and the local.

Let $G = (V, E)$ be an unweighted graph. The local clustering coefficient of a node in a graph quantifies how close its neighbours are to a clique. Formally, for a given node $v \in V$ of degree at least 2, the **local clustering coefficient** is defined as follows:

$$c(v) = \frac{|\{uw \in E : u, w \in N(v)\}|}{\binom{\deg(v)}{2}},$$

that is, $c(v)$ is the probability that two neighbours of v selected at random are adjacent. Alternatively, $c(v)$ can be defined as the ratio between the number of triangles involving node v and the number of paths of length two with v being its internal node. Clearly $0 \leq c(v) \leq 1$. The definition of $c(v)$ can be easily adjusted to directed graphs. For any node v in an unweighted directed graph $D = (V, E)$ that is of in-degree (respectively, out-degree) at least 2,

$$c^{in}(v) = \frac{|\{uw \in E : u, w \in N^{in}(v)\}|}{\deg^{in}(v) \cdot (\deg^{in}(v) - 1)}$$

$$c^{out}(v) = \frac{|\{uw \in E : u, w \in N^{out}(v)\}|}{\deg^{out}(v) \cdot (\deg^{out}(v) - 1)}.$$

(Note that in these definitions the order of nodes u and w matters.)

Note that the local clustering $c(v)$ is defined individually for each node and it can be noisy, especially for the nodes of not too large degrees. Therefore, the following global characteristic is usually studied. For $d \in \mathbb{N} \setminus \{1\}$, let $C(d)$ be the local clustering coefficient averaged over the nodes of degree d, that is,

$$C(d) = \frac{\sum_{v \in V : \deg(v) = d} c(v)}{|\{v \in V : \deg(v) = d\}|},$$

provided that $\{v \in V : \deg(v) = d\} \neq \emptyset$. $C(d)$ was extensively studied both theoretically and empirically. For example, it was observed for many real-world networks that $C(d)$ is proportional to $d^{-\varphi}$ for some $\varphi > 0$, often $\varphi \approx 1$.

The above definitions characterize a graph by a sequence $C(d)$ for various values of d. However, sometimes we are interested in a single number that gives us an aggregate information about the clustering of a graph. The **global clustering coefficient** $C_{\text{glob}}(G)$ of a graph G is the ratio of three times the number of triangles in G to the number of pairs of adjacent edges in G. In other words, if one samples a random pair of adjacent edges in G, then $C_{\text{glob}}(G)$ is the probability that these three nodes form a triangle. This measure gives an indication of the clustering in the whole network (global), and can be applied to both undirected and directed networks.

Alternatively, one may consider the average local clustering coefficient, that is, $C_{\text{loc}}(G) = \sum_{v \in V} c(v)/n$. Note that in the definition of $c(v)$ above we assumed that $\deg(v) \geq 2$. It is natural to extend this definition to isolated nodes and leaves (that is, when $\deg(v) \in \{0, 1\}$) by assigning $c(v) = 0$ to those nodes. Alternatively, one may define $C_{\text{loc}}(G)$ as the average over nodes of degree at least 2. Both variants are available in the `igraph` library and the latter one is assumed to be the default despite the fact that the former seems more natural. If G is d-regular, then $C_{\text{loc}}(G) = C_{\text{glob}}(G)$ but, in general, the two graph parameters can differ substantially. Consider, for example, the bike wheel graph that contains two hub nodes and n rim nodes. The hub nodes are connected (by the axel, in the bike terminology). Moreover, the hub nodes are connected to every rim node (by a spoke, in the bike terminology). No rim nodes are connected. Formally, the **bike wheel graph** is the graph $B_n = (V, E)$, where $V = \{a, b, 1, 2, \ldots, n\}$ and $uv \in E$ if and only if $\{u, v\} \cap \{a, b\} \neq \emptyset$. It is easy to see that

$$\lim_{n \to \infty} C_{\text{loc}}(B_n) = 1 \quad \text{whereas} \quad \lim_{n \to \infty} C_{\text{glob}}(B_n) = 0.$$

To see an example of the other extreme, consider the **lollipop** graph L_n consisting of a complete graph on $\lceil n^{2/3} \rceil$ nodes and a path of length n attached to one of the nodes of the complete graph. Again, it is easy to see that this time

$$\lim_{n \to \infty} C_{\text{loc}}(L_n) = 0 \quad \text{whereas} \quad \lim_{n \to \infty} C_{\text{glob}}(L_n) = 1.$$

1.12 Experiments

Looking at some basic statistics such as the degree distribution and the distribution of lengths of shortest paths, one can often identify huge differences between different types of graphs. This, in turn, may affect the running time or quality of various algorithms that the user might be interested in. As a result, looking at such statistics is typically the first step of exploratory analysis before any serious experiment is performed. In this section, we look at an example of a social-type graph, and a transportation-type network.

The first graph we consider is a GitHub[1] developers network from June 2019. This graph is unweighted and undirected with 37,700 nodes and 289,003 edges. Nodes are associated with developers who have starred at least 10 repositories and are of two types: web developers or machine learning developers. There is an edge between two developers if they follow at least one common repository.

The second graph we consider is the electric grid network of Europe from 2016 extracted by GridKit[2], which is built from OpenStreetMap[3]. It shows interconnections between different power stations and joints in the European high voltage electricity grid network. This graph is also undirected and we do not assign any weights to the edges, although we could for example consider the physical distance as edge weights. There are 13,844 nodes and 17,277 edges in this graph.

The main characteristics of those graphs are summarized in Table 1.1, with respect to degree distribution, diameter, connected components, and clustering coefficient. For the GitHub graph, since there are two types of nodes, we additionally consider the two subgraphs induced by only one type of nodes, that is, the web developers (web) nodes and the machine learning developers (ml) nodes.

There are some striking differences between the two graphs, not only in terms of the corresponding edge densities, but also degree distributions. For the GitHub graphs, the median degree is lower than the mean degree, and there are some very high degree nodes. As we will see later, this is a common property of graphs that model some social interactions: there are many low degree nodes and a few nodes of very high degree (see Section 2.4). For the grid graph, however, the maximum degree is much smaller so its degree distribution is more uniform. This graph also has a much larger diameter (147), so its nodes are not as tightly connected in comparison to the GitHub graph. Looking at the distribution of connected components, the grid network has one large (we usually call it giant) component and several (58) small ones, while the GitHub graph is connected. However, if we consider the GitHub induced subgraphs, we observe many components, mainly due to appearance of isolated nodes. Those are the result of 2,308 ml developers who are only linked to web developers, and 285 web developers that are only linked to ml developers.

In Figure 1.2, we illustrate some selected parts of each graph under study. When visualizing a graph, one needs to specify a *layout*, that is, the way nodes are placed in 2 or 3 dimensional space. For the GitHub graph, we used a force directed layout which is a type of graph embedding that will be discussed in Section 6.5. For the grid network, there is a natural embedding provided by the latitude and longitude of the power station associated with nodes. In both cases, since the original graphs are too large, we only display

---

[1] snap.stanford.edu/data/github-social.html
[2] zenodo.org/record/47317#.X1aWCCOZNhG
[3] www.openstreetmap.org

TABLE 1.1

Basic descriptive statistics for the GitHub and the Grid graphs. The GitHub subgraphs built with the two types of developers (ml and web) are also included. $d_{quant_{99}}$ refers to the 99th quantile for the degree distribution.

graph	GitHub	GitHub (ml)	GitHub (web)	Grid
# nodes	37,700	9,739	27,961	13,844
# edges	289,003	19,684	224,623	17,277
δ	1	0	0	1
$\langle k \rangle$	15.332	4.042	16.067	2.496
median degree	6	2	6	2
$d_{quant_{99}}$	138	39	145	8
Δ	9,458	482	8,194	16
diameter	11	13	9	147
# components	1	2,466	297	59
the largest component	37,700	7,083	27,653	13,478
# isolates	0	2,308	285	0
C_{glob}	0.012	0.034	0.014	0.100
C_{loc}	0.193	0.141	0.207	0.113

selected snapshots of the graphs. For the Grid network, we selected values that correspond roughly to the Iberian peninsula. On the other hand, for the GitHub graph, we selected a region in the center of the layout of the giant component respectively for both the web developer and ml developer subgraphs. The sizes of the regions of the graphs were selected so that we depict roughly 1,000 nodes in each picture. Given the statistics presented in Table 1.1, it is not surprising to see some dense regions in the GitHub subgraphs, with a mixture of nodes of high and low degrees, while the Grid network shows much more regular patterns in terms of the degree distribution.

Another way to look at the degree distribution is to plot the (empirical) cumulative distribution function (CDF), as we do in Figure 1.3. In such plots, one presents points $(\ell, d_{\leq \ell})$, where $d_{\leq \ell}$ is the fraction of nodes of degree at most ℓ. While the shapes of the corresponding curves look similar, we see that the GitHub degree CDF extends to very large values for the degrees, a phenomenon known as a "long tailed" distribution often present in social graphs that we will study in later chapters.

In Figure 1.4, we look at the distribution of the number of shortest paths of each length between nodes in each graph. An approximation of such distributions is achieved by taking a random sample of 100 nodes and computing the shortest path lengths (number of hops) to reach every other node in the graph. Here again, we see striking differences between the two graphs. For the GitHub graph, the shortest paths are all quite short, a phenomenon present in many real-world networks that we will refer to as "small world" phenomenon. On the other hand, in the grid networks some paths are very long. Indeed,

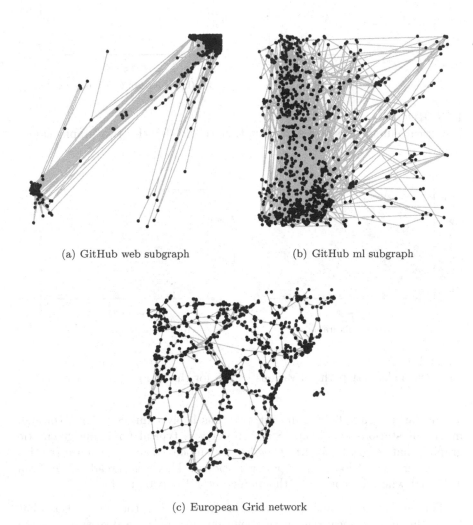

(a) GitHub web subgraph

(b) GitHub ml subgraph

(c) European Grid network

FIGURE 1.2
Snapshots of the graphs under study.

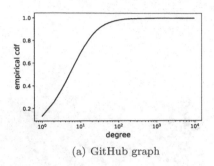
(a) GitHub graph

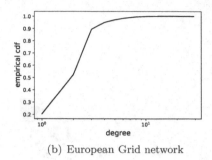
(b) European Grid network

FIGURE 1.3
Empirical cumulative degree distribution (CDF) for the two graphs under study.

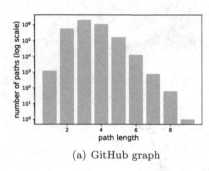
(a) GitHub graph

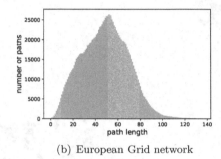
(b) European Grid network

FIGURE 1.4
Number of shortest paths of various length for the two graphs under study.

as one can imagine, linking distant locations likely requires passing through numerous stations and joints. Such property is typical for many geometric graphs, that is, graphs in which nodes are embedded in some space (in this case, geographical location of power stations). This was already clear from Table 1.1 where we compared the diameters of the two graphs.

For our last experiment, we experimentally verify the observation that $C(d)$, the average local clustering coefficient for nodes of degree d, is often proportional to $d^{-\varphi}$ for some $\varphi > 0$. We consider the GitHub developers graph. In Figure 1.5, we plot the values of $C(d)$ in the range $10 \leq d \leq 1,000$ using a log-log scale. We see that the expected linear relation is clear, in particular, for higher degree nodes. We also show the line we obtained via linear regression which had a slope of -0.716.

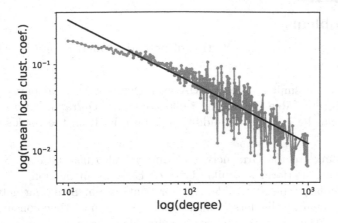

FIGURE 1.5
Comparing the mean local clustering coefficients for each degree d as a function of d for the GitHub graph. We also show the line obtained via linear regression in black.

1.13 Practitioner's Corner

When tackling some problems using graph mining tools, it may be the case that the graph is explicitly available to be used (such as the Grid network) while in the other case some pre-processing needs to be done to build the graph. Such pre-processing step needed to be done for the GitHub graph, where nodes correspond to a subset of developers that starred at least 10 repositories and edges were created between developers that follow at least one common repository.

While the entire small graphs can be easily visualized, it is often not the case for larger graphs. In that case, the usual first step in data exploration—known as EDA (exploratory data analysis)—is performed by looking at some basic statistics of the graph. As we saw in this chapter, simple statistics such as the degree distribution, the distribution of path lengths and connected components already give a good sense of a general topology of the graph at hand.

1.14   Problems

In this section, we present a collection of potential practical problems for the reader to attempt.

1. Perform a similar type of EDA as in Section 1.12 (in particular, regenerate Table 1.1) on the 1,000 node ABCD graph. Data (edge list) can be found in the additional material from the book's web site.

2. The link[4] to the Grid network (Europe) also has data for North America. Perform a similar type of EDA as in Section 1.12 (in particular, replicate Table 1.1) for that graph, including getting and preparing the data (which is usually the most time consuming part of the process). The raw downloaded files can be also found in `Datasets/GridNorthAmerica` in the additional material from the book's website. You may have a look at the raw files and the processed data in `Datasets/GridEurope` and do the same for North America.

3. Consider the GitHub (ml) graph on 9,739 nodes.

    a. Find the number of walks of length 5.

    b. Find the number of cycles of length 4 (induced, that is, without chords).

4. Consider the airport graph[5] found in the additional material from the book's web site under `Datasets/Airport/airport_data.csv`. The first 3 fields are: the origin, the destination and, respectively, the number of passengers. This graph is weighted and directed; the weight of a directed edge uv corresponds to the number of passengers travelling from airport u to airport v.

    a. Plot the cumulative degree distribution (points $(\ell, n_{\geq \ell})$, where $n_{\geq \ell}$ is the number of nodes of degree at least $(\ell \geq 1)$ in the log-log plot. Find the slope of the line obtained via linear regression.

    b. Find the busiest airport. In other words, find the node with the maximum total degree $(\deg^{in}(v) + \deg^{out}(v))$.

    c. Find the number of strongly connected components.

    d. Find the subgraph induced by all airports from California. Find the number of isolated nodes in this subgraph.

---

[4]`zenodo.org/record/47317#.X1aWCCOZNhG`
[5]built from `www.kaggle.com/flashgordon/usa-airport-dataset`

5. Take 100 random pairs of nodes in the European Grid network. For each pair plot a point (x, y), where x is the graph distance between the two nodes and y is the corresponding geographical distance (in kilometres). Is there a correlation between the two distances? In order to compute the geographical distance you may, for example, use function `geodesic` from `geopy` package or implement it from scratch using the haversine formula that determines the great-circle distance between two points on a sphere given their longitudes and latitudes.

6. For a given n, we generate a graph on the set of nodes $[n]$ as follows. We randomly pick a set $S \subseteq [n]$ of size $\lceil n/10 \rceil$. We put an edge between u and v with probability $1/2$, provided that at least one of the nodes is outside of S; otherwise, there is no edge between u and v. As a result, we get a graph with an independent set of size at least $n/10$. Finding it for large n is challenging!

   In order to find independent sets, you can try to use a standard function in `igraph`. Alternatively, you may write a short piece of code and greedily generate one independent set. Start with an empty set T and consider nodes in some order. Add node v to T if it is not adjacent to any node already in T.

   a. Implement an algorithm generating the random graph we introduced above (on n nodes and independent set of size at least $n/10$). For comparison purposes, we provided a piece of code at the end of the notebook for Chapter 2.

   b. Implement the greedy algorithm for finding an independent set (not necessarily one of the largest size) that we introduced above.

   c. Create 11 random networks on $n \in \{100, 200, \ldots, 100 \times 2^k$ for $0 \le k \le 10\}$ nodes.

   d. Try to run the standard function from `igraph` on the networks generated in part c. (It might take too long for large graphs so give up once you get bored waiting.) Report the size of the independent set found and the running time.

   e. Run the greedy algorithm on the networks generated in part c. Report the size of the independent set found and the running time.

7. Recall that there are two variants of the average local clustering coefficient (one of them ignores nodes of degree 0 or 1; the other one assigns $c(v) = 0$ to these nodes). Both variants are available in `igraph`. Find both values for the graphs presented in Table 1.1. (The default variant is already reported there.)

1.15   Recommended Supplementary Reading

There are many books on graph theory, mostly focusing on theoretical aspects. We will introduce all necessary definitions and results we need for our purpose but the reader interested in theory is encouraged to look at the following books.

- *Introduction to Graph Theory*, Douglas B. West, 2nd Edition, Pearson, Upper Saddle River: Prentice hall, 2000.

- *Graph Theory*, Reinhard Diestel, 5th Edition, Springer, Springer-Verlag Berlin Heidelberg, 2017.

- *Modern Graph Theory*, Béla Bollobás, 2nd Edition, Springer, Springer Science & Business Media, 2002.

If needed, in order to refresh basic knowledge in linear algebra, we recommend to look at one of the many textbooks on the topic including the following two positions.

- *Introduction to Linear Algebra*, Gilbert Strang, 5th Edition, Wellesley-Cambridge Press, Wellesley, MA, 2016.

- *Introduction to Applied Linear Algebra*, Stephen Boyd, Lieven Vandenberghe, Cambridge University Press, 2018.

It might be good to look at the documentation of the libraries used in the accompanying notebooks.

- `igraph` library, API Documentation[6].

- `LightGraphs` library, used in the Julia examples[7].

We introduced three datasets in this chapter. Here are the references for, respectively, the GitHub developers network dataset and the SNAP repository where it is hosted, the Grid dataset and the Kaggle repository for the US Airport dataset.

- B. Rozemberczki, C. Allen, and R. Sarkar, *Multi-scale Attributed Node Embedding*, arXiv (2019). https://arxiv.org/abs/1909.13021.

- J. Leskovec and A. Krevl, *SNAP Datasets: Stanford Large Network Dataset Collection*, http://snap.stanford.edu/data, 2014.

- B. Wiegmans, *GridKit: European and North-American extracts*, 2016, Zenodo. http://doi.org/10.5281/zenodo.47317

---

[6]`igraph.org/python`
[7]`github.com/JuliaGraphs/LightGraphs.jl`

- *Kaggle Public Datasets*: https://www.kaggle.com/datasets

Finally, as promised in the preface, we list a few books that also include topics related to mining networks.

- *A First Course in Network Science*, Filippo Menczer, Santo Fortunato, Clayton A. Davis, Cambridge University Press, 2020.

- *Network Science*, Albert-Lszl Barabsi, Cambridge University Press, 2016.

- *Networks*, Mark Newman, Oxford University Press, 2nd Edition, 2018.

- *Complex Networks: Principles, Methods and Applications*, Vito Latora, Vincenzo Nicosia, Giovanni Russo, Cambridge University Press, 2017.

2

Random Graph Models

2.1 Introduction

The theory of random graphs lies at the intersection between graph theory—which is part of the larger field of combinatorics—and probability theory. "Random graph" is a very general term that refers to probability distributions over some family of graphs. Random graphs may be described simply by providing an explicit probability distribution or by some random process which generates them.

There are many reasons why researchers and practitioners are interested in random graphs. They typically include the following four reasons (which are interrelated):

(i) They are interesting and surprising mathematical objects that can be used to answer questions about the properties of typical graphs. They are often used to support conjectures but may sometimes serve as counterexamples. Many models incorporate additional aspects such as geometry. This is a very active theoretical research field with many open questions still waiting to be answered.

(ii) They can be used to model real-world complex networks which, in turn, provide us with a better understanding of the underlying mechanisms that create them. A famous example of this research direction is the **Barabási–Albert (BA)** model (also known as the **preferential attachment** model), which generates a power-law network using the so-called "rich get richer" mechanism. It does not create a network that mimics real-world networks too closely, but it is a very important model as it shows that growth and preferential attachment mechanism are the main reasons why power-law graphs appear in many natural scenarios.

(iii) They can be used to create synthetic networks that closely resemble the real-world graphs that they attempt to model. Clearly, the process of generating synthetic graphs is easier than the process of cleaning and preparing a real-world network for experiments. Moreover, because of their flexibility, they provide a good tool for testing various scenarios. For example, one might need to understand how a certain algorithm running on, say, the Facebook

DOI: 10.1201/9781003218869-2

graph is going to perform if the number of users doubles or the average number of friends triples. A few such synthetic benchmarks are discussed in Section 5.3.

(iv) They can be used to benchmark the outcome of some algorithm of interest. For example, we might want to know whether some group of Facebook users form a community by analyzing the number of connections between them. If the number of friendships captured within that group is larger than what one would normally expect, then the answer is yes. But what should one expect? Random graphs provide an answer to that question. We discuss this particular application of random graphs in Section 5.4 but there are many other situations where random graphs are used. We will see them quite often in this book.

Since the material is selected with practitioners that are more interested in mining networks rather than modelling them in mind, we concentrate on (iii) and (iv). We present some basic models that try to preserve the average degree or the degree distribution. More advanced models will be introduced later in this book. However, this is not to say that (i) and (ii) are not interesting nor important.

This chapter is structured as follows: we first introduce the asymptotic notation that will be used in this chapter (Section 2.2). Next, we introduce the binomial random graph model and closely related models (Section 2.3). Unfortunately, many real-world networks exhibit different degree distributions than the one produced by these models (Section 2.4) so one needs to generalize them to allow for more flexibility. The Chung-Lu model provides such flexibility (Section 2.5). However, this model generates a random graph with only the *expected* degree distribution matching the desired one. The next two models generate random graphs whose degree distribution is an *exact* match for what we prescribe. We start with random d-regular graphs (Section 2.6) and then generalize the model to any degree distribution (Section 2.7). As usual, we finish the chapter with experiments (Section 2.8) and provide some tips for practitioners (Section 2.9).

2.2   Asymptotic Notation

Before we define various models of random graphs, let us stress the fact that none of these graphs are deterministic but are generated by some random process. In particular, as will be shown in the next section, the empty graph on n nodes can occur as $\mathcal{G}(n, 1/2)$ with exactly the same probability as the complete graph. However, both situations are extremely rare, provided that n is large enough. Hence, we will pay attention to properties that are typical for a given random model assuming that n is large. Formally, we say that an event in a given probability space holds **asymptotically almost surely** (**a.a.s.**), if its probability tends to one as n goes to infinity. For example, a.a.s. $\mathcal{G}(n, 1/2)$ is not the complete graph K_n nor the empty graph $\overline{K_n}$.

We will use the following **asymptotic notation**. Given two functions $f = f(n)$ and $g = g(n)$, we will write $f(n) = O(g(n))$ if there exists an absolute constant $c > 0$ such that $|f(n)| \leq c|g(n)|$ for all n, $f = \Omega(g)$ if $g = O(f)$, $f(n) = \Theta(g(n))$ if $f(n) = O(g(n))$ and $f(n) = \Omega(g(n))$, and we write $f(n) = o(g(n))$ or $f(n) \ll g(n)$ if the limit $\lim_{n\to\infty} f(n)/g(n) = 0$. In addition, we write $f(n) = \omega(g(n))$ or $f(n) \gg g(n)$ if $g(n) = o(f(n))$, and unless otherwise specified, $\omega = \omega(n)$ will denote an arbitrary function that is $\omega(1)$, and typically it is assumed to grow slowly as $n \to \infty$. We will also write $f(n) \sim g(n)$ if $f(n) = (1 + o(1))g(n)$.

In order to investigate the asymptotic behaviour of random graphs, we will use some standard inequalities and approximations. Let us summarize them below. The Taylor series for the exponential function e^x at $a = 0$

$$e^x = 1 + x + \frac{x^2}{2!} + \frac{x^3}{3!} + \ldots = 1 + x + O(x^2),$$

implies that $1 + x = \exp(x + O(x^2))$. For the factorial function $n!$, we have that for any $n \in \mathbb{N}$,

$$\left(\frac{n}{e}\right)^n \leq n! \leq en\left(\frac{n}{e}\right)^n.$$

For the asymptotic behaviour one can use the well-known **Stirlings formula**

$$n! \sim \sqrt{2\pi n}\left(\frac{n}{e}\right)^n.$$

For the binomial coefficient, we have that for any integers ℓ, n such that $0 \leq \ell \leq n$,

$$\left(\frac{n}{\ell}\right)^\ell \leq \binom{n}{\ell} \leq \left(\frac{en}{\ell}\right)^\ell.$$

Through this chapter—as is typical in the field of random graphs—expressions that clearly have to be an integer will be rounded up or down. We do not specify the way we do it as this choice does not affect the argument.

2.3 Binomial Random Graphs

Let us start with definitions of three models that are commonly used. Note that we typically use V for the set of nodes but for random graph models we make an exception and use $[n]$ instead. The reason for this is two-fold: first of all, we want to stress the fact that we work with labelled graphs and labelling nodes with consecutive natural numbers seems natural. The second reason is that most results are asymptotic in nature, that is, when the number of nodes n grows. As a result, the model actually introduces an infinite family of graphs (the sequence), not just a single graph on nodes from a fixed set V.

> The **binomial random graph** $\mathcal{G}(n,p)$ can be generated by starting with the empty graph on the set of nodes $[n] = \{1, 2, \ldots, n\}$. For each pair of nodes i, j such that $1 \leq i < j \leq n$, we independently introduce an edge ij in G with probability p.

Note that $p = p(n)$ may (and usually does) tend to zero as n tends to infinity. Let us also note that we formally defined the probability distribution over a family of labelled graphs on n nodes. There are $\binom{N}{m}$ labelled graphs on n nodes and m edges, where $N = \binom{n}{2}$ is the number of pairs of nodes. For a given labelled graph G on n nodes and m edges, we obtain this graph with probability

$$\mathbb{P}(G) = p^m (1-p)^{N-m}.$$

Hence, indeed, the probability is well-defined as

$$\sum_{m=0}^{N} \binom{N}{m} p^m (1-p)^{N-m} = 1,$$

by the binomial theorem. Let us mention that $p = 1/2$ plays a special role in this model as in this case $\mathbb{P}(G) = 2^{-N}$ for any graph G on n nodes, regardless of how many edges G has. As a result, the corresponding probability space is the uniform probability space over the family of all graphs on n nodes.

The second model we would like to mention is the uniform random graph $\mathcal{G}(n,m)$. The two models, $\mathcal{G}(n,p)$ and $\mathcal{G}(n,m)$, are in many cases asymptotically equivalent, provided $\binom{n}{2}p$, the expected number of edges in $\mathcal{G}(n,p)$, is close to m.

> Let Ω be the family of all labelled graphs on the set of nodes $[n]$ and exactly m edges, where $0 \leq m \leq N$, $N = \binom{n}{2}$. The **uniform random graph** $\mathcal{G}(n,m)$ assigns to every graph $G \in \Omega$ the same probability, that is,
>
> $$\mathbb{P}(G) = \frac{1}{|\Omega|} = \binom{N}{m}^{-1}.$$

Alternatively, the uniform random graph $\mathcal{G}(n,m)$ can be generated using the following random process:

The **Erdős-Rényi random graph process** is a stochastic process that starts with n labelled nodes and no edges, and at each step adds one new edge chosen uniformly at random from the set of missing edges. Formally, let $N = \binom{n}{2}$ and let $e_1, e_2, \ldots, e_N$ be a random permutation of the edges of the complete graph K_n. The graph process consists of the sequence of random graphs $(\mathcal{G}(n, m))_{m=0}^N$, where $\mathcal{G}(n, m) = ([n], E_m)$ and $E_m = \{e_1, e_2, \ldots, e_m\}$. It is clear that $\mathcal{G}(n, m)$ is a graph taken uniformly at random from the set of all graphs on n nodes and m edges.

Let us now present a few results from the theory of random graphs. This is a very active field so we can only scratch the surface. Perhaps the most studied phenomenon in the field of random graphs is the behaviour of the binomial random graph when its average degree $\langle k \rangle$ is near 1. Though we will distinguish 3 phases in its evolutionary process, the big picture is actually more complex and interesting. The **sub-critical phase** happens when $\langle k \rangle < 1 - \epsilon$ for some $\epsilon > 0$. During that phase, a.a.s. $\mathcal{G}(n, p)$ consists of small trees and unicyclic components, and thus its structure is rather easy to study. The size of the largest component is of order $\ln n = o(n)$ and so it is of negligible size. A **giant component** (that is, a connected component of linear size) is formed from smaller ones during the so-called **critical phase** when $\langle k \rangle \sim 1$. During that phase, the size of the largest component keeps growing, reaching $\Theta(n^{2/3})$ nodes at precisely $\langle k \rangle = 1$ a.a.s. After that, a.a.s. $\mathcal{G}(n, p)$ consists of one complex component (that is, a connected component with more edges than nodes) of growing size and some number of small trees and unicyclic components. However, the size of the largest component and the number of edges in this component are still $o(n)$, provided that $\langle k \rangle \sim 1$. Moreover, the size of the second-largest component keeps decreasing. Finally, when $\langle k \rangle > 1 + \epsilon$ for some $\epsilon > 0$, a.a.s. the size of the giant component is $(1 + o(1))\beta n$, where $\beta = \beta(\langle k \rangle)$ is the positive real number satisfying

$$\beta + e^{-\beta \cdot \langle k \rangle} = 1.$$

The giant component is unique and the second-largest component is acyclic or unicyclic and has the size of order $\ln n$. This phase is called the **very super-critical phase**.

These theoretical and asymptotic formulas are compared with simulation results for small graphs in Figure 2.1. We performed 1,000 independent runs with a wide range of values for the average degree. For better readability, we coupled the graphs so that $\mathcal{G}(n, p_1)$ is a subgraph of $\mathcal{G}(n, p_2)$ whenever $p_1 < p_2$. The 90% confidence interval is presented as the shaded area. The experiments for small graphs on $n = 100$ nodes show some variability but the results are quite close to the corresponding expected values. For larger graphs on $n = 10,000$ nodes, the empirical results agree almost perfectly with the theoretical predictions.

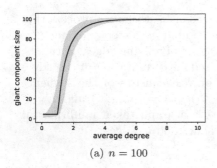
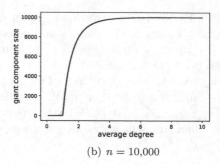

(a) $n = 100$ (b) $n = 10,000$

FIGURE 2.1
The order of the giant component: theoretical predictions and empirical results based on 1,000 independent runs for small graphs on $n = 100$ nodes (a) and larger graphs on $n = 10,000$ (b) nodes.

Another important phenomenon is the threshold for connectivity. Suppose that

$$p = p(n) = \frac{\ln n + c}{n}$$

for some constant $c \in \mathbb{R}$. Note that the expected number of isolated nodes is equal to

$$
\begin{aligned}
n\,(1-p)^{n-1} &= n\exp\left(-\frac{\ln n + c}{n} + O\left(\frac{(\ln n)^2}{n^2}\right)\right)^{n-1} \\
&= n\exp\left(-(\ln n + c) + o(1)\right) \sim e^{-c}.
\end{aligned}
$$

If c is negative and small (that is, $|c|$ is large and $c < 0$), then we expect to see a lot of isolated nodes. On the other hand, if c is positive and large, then we expect the opposite. Clearly, if there are isolated nodes, then the graph is disconnected though there are many disconnected graphs without any isolated nodes. However, it is possible to show that this trivial necessary condition for connectivity is in fact the main obstacle for $\mathcal{G}(n,p)$ to be connected. Once isolated nodes disappear, the binomial random graph is connected a.a.s. The following property holds:

$$
\mathbb{P}\Big(\mathcal{G}(n,p)\text{ is connected}\Big) \sim
\begin{cases}
0 & \text{if } c \to -\infty \\
e^{-e^{-c}} & \text{if } c \in \mathbb{R} \\
1 & \text{if } c \to \infty.
\end{cases}
$$

In particular, we get the following corollary of this much stronger result. If $pn < (1-\epsilon)\ln n$ for some $\epsilon > 0$, then a.a.s. $\mathcal{G}(n,p)$ is disconnected. On the other hand, if $pn > (1+\epsilon)\ln n$ for some $\epsilon > 0$, then a.a.s. $\mathcal{G}(n,p)$ is connected.

As before, in Figure 2.2, the asymptotic probability that the random graph $\mathcal{G}(n,p)$ with $np = \ln n + c$ is connected (that is, $f(c) = e^{-e^{-c}}$) is compared with simulation results for graphs of order $n = 100$ and $n = 10{,}000$. As before, in order to obtain better readability, we coupled the graphs so that $\mathcal{G}(n, p_1)$ is a subgraph of $\mathcal{G}(n, p_2)$ whenever $p_1 < p_2$. The shaded areas correspond to the 90% confidence intervals for the estimates obtained with 1,000 runs, and the black lines correspond to the theoretical values $f(c)$. Despite some variability, the results are quite close to the expected values.

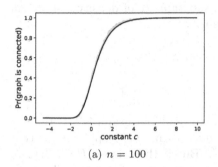
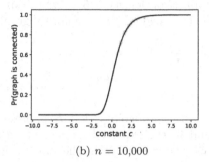

(a) $n = 100$ (b) $n = 10{,}000$

FIGURE 2.2
The probability that $\mathcal{G}(n,p)$ with $np = \ln n + c$ is connected: theoretical predictions and empirical estimations based on 1,000 independent runs for small graphs on $n = 100$ (a) nodes and larger graphs on $n = 10{,}000$ nodes (b).

One important local property of networks are so-called network **motifs**, which are defined as small subgraphs of a large network. Depending on the application at hand, it is sometimes natural to insist that these small subgraphs are induced (see Section 1.9 for the definition). For random graphs, non-edges do not cause any problems, unless p is close to 1 but this is usually not an interesting case anyway. Hence, in order to keep things as simple as possible, we will concentrate on subgraphs, not necessarily induced.

Before we state the main observation, let us consider the following illustrative example. We would like to understand if the graph G presented in Figure 2.3 appears in $\mathcal{G}(n,p)$ with $p = p(n) = n^{-9/11}$ as a subgraph or not. Note that the expected number of copies of G in $\mathcal{G}(n,p)$ is equal to

$$\binom{n}{5} \frac{5!}{2} p^6 = \Theta(n^5 p^6) = \Theta(n^{1/11}) \to \infty.$$

Indeed, there are $\binom{n}{5}$ choices for the 5 nodes of G, $5!/2$ possible ways to embed G on those 5 nodes, and after that edges appear at the selected places with probability p^6. Hence, one might think that there should be at least one copy of G present in the random graph. However, we get the opposite intuition if we consider a subgraph H of G. The expected number of copies of H in $\mathcal{G}(n,p)$

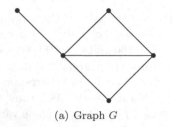
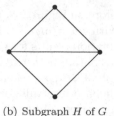

(a) Graph G (b) Subgraph H of G

FIGURE 2.3
Graph G of density $6/5 = 1.2$ (a) and its subgraph H of density $5/4 = 1.25$
(b) that yields the maximum subgraph density.

is equal to

$$\binom{n}{4} \frac{4!}{4} p^5 = \Theta(n^4 p^5) = \Theta(n^{-1/11}) \to 0.$$

Since the probability that H is present in the random graph is less than or
equal to the expected number of its occurrences in the graph, we get that
a.a.s. there is no H in the random graph. But if there is no H in $\mathcal{G}(n,p)$,
then there is no G! How can one explain this paradox? Note that if we see
a single copy of H, then we expect to see $2(n-4)p = \Theta(n^{2/11})$ copies of G
built around the copy of H. So we expect to see it $\Theta(n^{1/11})$ more times than
we expect to see G unconditionally. In summary, the probability that there is
at least one copy of G (and so also at least one copy of H) in $\mathcal{G}(n,p)$ tends
to 0 as $n \to \infty$, but conditioning on the fact that there is at least one copy of
G in the random graph, the expected number of copies of G tends very fast
to infinity as $n \to \infty$. In general, one can show that the maximum subgraph
density of G plays a more important role than the density of G in this context.

Let us formalize these observations. For a given motif $G = (V, E)$, let
$d(G) = |E|/|V|$ be the **density** of G, and let the **maximum subgraph
density** of G be defined as follows:

$$m(G) = \max\{d(H) : H \subseteq G\},$$

where the maximum is taken over all subgraphs H of G. If $pn^{1/m(G)} \to 0$,
then a.a.s. $\mathcal{G}(n,p)$ does not contain G as a subgraph. On the other hand, if
$pn^{1/m(G)} \to \infty$, then a.a.s. $\mathcal{G}(n,p)$ contains G as a subgraph. The function $p =
n^{-1/m(G)}$ is called the **threshold subgraph probability** for the property
that $\mathcal{G}(n,p)$ contains G as a subgraph.

Recall that the diameter of a connected graph $G = (V, E)$ is defined as
follows: $\mathrm{diam}(D) = \max_{u,v \in V} \mathrm{dist}(u,v)$, where $\mathrm{dist}(u,v)$ is the distance between nodes u and v. Since the number of nodes at distance ℓ from any node
v is expected to be around $\min\{\langle k \rangle^\ell, n\}$, we expect the diameter of the giant
component of $\mathcal{G}(n,p)$ to be close to $\ln n / \ln\langle k \rangle$. This is true provided that the
average degree is large enough but might be slightly off for sparse graphs. In

particular, the following property holds. Suppose that $\langle k \rangle = (n-1)\, p \gg \ln n$ and

$$\langle k \rangle^i / n - 2\ln n \to \infty \qquad \text{and} \qquad \langle k \rangle^{i-1}/n - 2\ln n \to -\infty.$$

Then, the diameter of $\mathcal{G}(n,p)$ is equal to i a.a.s.

Similar observations can be made for the average distance between two nodes from the giant component of $\mathcal{G}(n,p)$. We present the results of one experiment with binomial random graphs with expected degree 5 in Figure 2.4.

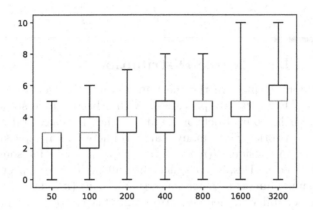

FIGURE 2.4
Distribution of shortest path lengths for the giant component of binomial random graphs with expected degree 5 and varying number of nodes. The extreme value for each boxplot corresponds to the diameter. We see a slow increase in the diameter as we double the number of nodes between each boxplot.

Finally, let us turn our attention to algorithmic aspects. In order to generate a sparse uniform random graph $\mathcal{G}(n,m)$, one typically starts from an empty graph $G = ([n], E)$ with $E = \emptyset$. Next, we repeatedly sample nodes u and v from $[n]$ independently and uniformly and if $u \neq v$, then we add an edge uv to E. Note that uv might already be present in E in which case uv is discarded, that is, E is not changed. The process ends when $|E| = m$. For dense graphs (that is, when $m > \binom{n}{2}/2$), a standard procedure is to sample the graph $\mathcal{G}(n, \binom{n}{2} - m)$ and then take its complement, which has the same distribution as $\mathcal{G}(n,m)$.

Now, we briefly analyze how long this process takes, provided that $m < \binom{n}{2}/2$. Assuming that m' edges are already present, the expected number of samples that we need to perform to generate a new edge is equal to

$$\frac{\binom{n}{2}}{\binom{n}{2} - m'} = \frac{n(n-1)}{n(n-1) - 2m'} \leq \frac{n(n-1)}{n(n-1) - 2m} \leq 2.$$

So in total, we only need $(1 + o(1))m$ samples, provided that $m = o(n^2)$ and at most $2m$ for dense graphs. Hence, the generation process is very fast, one can do it in linear time as a function of m.

In order to generate a binomial random graph $\mathcal{G}(n,p)$, we first sample the number of edges m from $\text{Bin}(\binom{n}{2}, p)$ distribution and next use the $\mathcal{G}(n,m)$ sampler described above. The efficiency of a method to sample from $\text{Bin}(\binom{n}{2}, p)$ is dependent upon how large the value of its expectation is. It is implemented in this form in Python, Julia, and most other computing environments.

2.4   Power-Law Degree Distribution

Binomial random graphs are the most important and well-studied random graph models. They certainly provide us with a better understanding of the formation and behaviour of networks but can they be successfully applied as models of real networks? Since many real-world networks are sparse and large, we might want to consider $\mathcal{G}(n,p)$ with $p = p(n) = c/n$ for some constant $c \in \mathbb{R}_+$ and large n. Despite the fact that all nodes are expected to have exactly the same number of neighbours (namely, $p \cdot (n-1) \sim c$), there will still be nodes of different degrees. After all, this is not a c-regular graph. Indeed, for any node $v \in [n]$ and $\ell \in \mathbb{N} \cup \{0\}$,

$$
\begin{aligned}
\mathbb{P}\Big(\deg(v) = \ell \Big) &= \binom{n-1}{\ell} p^\ell (1-p)^{n-1-\ell} \\
&= \frac{(n-1)\cdots(n-\ell)}{\ell!} \left(\frac{c}{n}\right)^\ell \exp\left(-\frac{c}{n} + O\left(n^{-2}\right)\right)^{n-1-\ell} \\
&\sim \frac{n^\ell}{\ell!} \frac{c^\ell}{n^\ell} \exp\left(-c + o(1)\right) \sim \frac{c^\ell}{\ell!} e^{-c}.
\end{aligned}
$$

Moreover, the corresponding events are asymptotically independent. For any two nodes, u, v, and $\ell_1, \ell_2 \in \mathbb{N} \cup \{0\}$,

$$
\mathbb{P}\Big(\deg(v) = \ell_1 \wedge \deg(u) = \ell_2 \Big) \sim \mathbb{P}\Big(\deg(v) = \ell_1 \Big) \mathbb{P}\Big(\deg(u) = \ell_2 \Big).
$$

Hence, in the limit, the degree distribution of sparse $\mathcal{G}(n,p)$ can be approximated by the **Poisson distribution**, that is,

$$
d_\ell \sim \frac{c^\ell}{\ell!} e^{-c},
$$

where c is the asymptotic expected average degree. (Recall that d_ℓ is the fraction of nodes of degree ℓ.) We illustrate this in Figure 2.5 for a random graph with $n = 10,000$ nodes and $np = 10$. In particular, it follows that the number of nodes of degree ℓ decreases very fast as ℓ increases. As a result, the maximum degree is relatively small. Indeed,

$$\mathbb{P}\left(\deg(v) \geq \ell\right) \ \leq \ \binom{n}{\ell}p^\ell \leq \left(\frac{en}{\ell} \cdot \frac{c}{n}\right)^\ell = \exp\left(\ell \ln\left(\frac{ec}{\ell}\right)\right)$$

$$\leq \ \exp\left(-\ell(\ln\ell - \ln(ec))\right) = o(1/n),$$

provided that, say, $\ell = 2\ln n/\ln\ln n$ and n is large enough. It follows that a.a.s. the maximum degree is $O(\ln n/\ln\ln n)$.

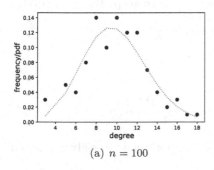 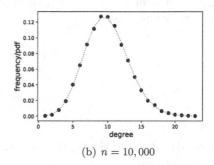

(a) $n = 100$ (b) $n = 10,000$

FIGURE 2.5
Degree distribution of $\mathcal{G}(n,p)$ with $n = 100$ (a) and $n = 10,000$, (b) and with $np = 10$. The dashed line is the corresponding Poisson distribution with $c = 10$.

Unfortunately, this means that binomial random graphs do not accurately reflect the degree distribution of most real-world networks. For example, Cristiano Ronaldo and Ariana Grande are considered to be the most popular users on Instagram (May 2020) because of a large number of followers: 216M+ and 183M+, respectively. Hence, there are large degree nodes in such networks, something that is extremely unlikely to see in $\mathcal{G}(n,p)$. After a number of experiments on real-world graphs, it was found that their degree distribution produced an approximately straight line on a log-log scale. This implies that the degree distribution is well approximated by the following equation:

$$d_\ell \approx c \cdot \ell^{-\gamma} \tag{2.1}$$

for some parameter $\gamma > 0$ and normalizing constant $c > 0$. Indeed, after taking a logarithm of both sides we get

$$\ln d_\ell \approx -\gamma \ln\ell + \ln c,$$

which yields a straight line with the slope $-\gamma$. Equation (2.1) is called a **power-law distribution** and the exponent γ is its **degree exponent**. The frequency of the natural occurrence of this distribution in practice was first observed by Vilfredo Pareto, a 19th-century economist, who observed that a few wealthy individuals posses the majority of world wealth. Finally, let us mention that the same observations apply to directed networks. For example, for the web graph, both in-degree and out-degree distributions follow the power-law distribution but with different degree exponents, namely, $\gamma_{in} \approx 2.1$ and $\gamma_{out} \approx 2.45$, which are approximations of the corresponding degree exponents that are typically reported.

Dealing with power-law distributions of discrete variables brings some technical challenges. Suppose that we want to generate a degree distribution with degree exponent $\gamma > 1$ and minimum degree $\delta \geq 1$. To keep things a little bit simpler, assume that

$$d_\ell = c \int_\ell^{\ell+1} x^{-\gamma} dx \approx c \cdot \ell^{-\gamma}.$$

Note that

$$1 = \sum_{\ell \geq \delta} d_\ell = c \int_\delta^\infty x^{-\gamma} dx = \frac{c\left(-\delta^{1-\gamma}\right)}{1-\gamma},$$

provided that $\gamma > 1$. Hence, the normalizing constant c can be easily computed and we get that

$$d_\ell \approx (\gamma - 1)\, \delta^{\gamma-1}\, \ell^{-\gamma}.$$

The average degree of a graph with this distribution is equal to

$$\begin{aligned}
\langle k \rangle &= \sum_{\ell=\delta}^{\Delta} \ell\, d_\ell \approx (\gamma - 1)\, \delta^{\gamma-1} \int_\delta^\infty x^{1-\gamma} dx \\
&= (\gamma - 1)\, \delta^{\gamma-1} \frac{-\delta^{2-\gamma}}{2-\gamma} = \frac{\gamma-1}{\gamma-2}\, \delta,
\end{aligned} \tag{2.2}$$

provided that $\gamma > 2$. In a few places in this book, we will also need the second moment, which is equal to

$$\begin{aligned}
\langle k^2 \rangle &= \sum_{\ell=\delta}^{\Delta} \ell^2\, d_\ell \approx (\gamma - 1)\, \delta^{\gamma-1} \int_\delta^\infty x^{2-\gamma} dx \\
&= (\gamma - 1)\, \delta^{\gamma-1} \frac{-\delta^{3-\gamma}}{3-\gamma} = \frac{\gamma-1}{\gamma-3}\, \delta^2,
\end{aligned} \tag{2.3}$$

provided that $\gamma > 3$.

Let us again stress that the average degree $\langle k \rangle$ is finite and well defined only for $\gamma > 2$ and the second moment $\langle k^2 \rangle$ only for $\gamma > 3$; otherwise, they tend to infinity as $n \to \infty$. Indeed, assuming that $\Delta \gg \delta$, it follows from (2.2) that

$$\langle k \rangle \approx \begin{cases} \frac{\gamma-1}{2-\gamma}\, \delta^{\gamma-1}\, \Delta^{2-\gamma} & \text{for } \gamma \in (1, 2) \\ (\gamma - 1)\, \delta^{\gamma-1} \ln(\Delta/\delta) & \text{for } \gamma = 2. \end{cases}$$

Similarly, from (2.3) we get that

$$\langle k^2 \rangle \approx \begin{cases} \frac{\gamma-1}{3-\gamma}\,\delta^{\gamma-1}\,\Delta^{3-\gamma} & \text{for } \gamma \in (2,3) \\ (\gamma-1)\,\delta^{\gamma-1}\ln(\Delta/\delta) & \text{for } \gamma = 3. \end{cases}$$

Since many real-world networks exhibit power-law degree distribution with exponent $\gamma \in (2,3)$, the average degree $\langle k \rangle$ in such graphs is finite and well defined whereas the second moment $\langle k^2 \rangle$ and the maximum degree Δ grow together with n. It is possible to estimate how Δ depends on n by investigating the following condition

$$1 = n \sum_{\ell \geq \Delta} d_\ell = n\,(\gamma-1)\,\delta^{\gamma-1} \int_\Delta^\infty x^{-\gamma} dx \approx n\,(\delta/\Delta)^{\gamma-1},$$

provided $\gamma > 1$. This condition simply assumes that the number of nodes of degree at least Δ is close to 1. We get that

$$\Delta = \delta\,n^{1/(\gamma-1)}, \tag{2.4}$$

which is often referred to as the **natural cut-off** of the graph.

Finally, let us mention that there is another upper bound for the maximum degree in power-law graphs, namely, the so-called **structural cut-off**. The reason for cutting the high degree nodes is that one expects a large number of edges between high degree nodes that are present in power-law graphs. However, as we deal with simple graphs, there is simply not enough room for such edges which creates a problem. Indeed, it can be shown that not all distributions d_ℓ can be achieved if one is restricted to a family of simple graphs. Recall that a sequence of numbers is said to be a **graphic sequence** if one can construct a graph having the sequence as its degree sequence. Assuming that the sum of degrees of all nodes present in the graph is even, the following condition is both sufficient and necessary for the degree sequence to be graphic: for all $r \in [n-1]$,

$$\sum_{i=1}^{r} \deg(v_i) \leq r(r-1) + \sum_{i=r+1}^{n} \min\{r, \deg(v_i)\}, \tag{2.5}$$

where v_i is a permutation of nodes from V that makes the sequence non-increasing, that is, $\deg(v_i) \geq \deg(v_{i+1})$ for all $i \in [n-1]$. So, indeed, we see that the degree distribution d_ℓ cannot allow for a few nodes with very large degree and a large number of nodes with small degrees. We discuss this issue in more detail for special graph classes in Section 4.4.

2.5   Chung-Lu Model

As mentioned in the previous section, binomial random graphs are not flexible enough to mimic degree distributions of most real-world networks. In order to overcome this problem, one can use the Chung-Lu model which produces random graphs with an expected degree sequence following a given sequence $\mathbf{w}$, provided that we accept the generated graph to be non-simple (that is, it may contain self loops or multiple edges).

Let $\mathbf{w} = (w_1, w_2, \ldots, w_n)$ be any vector of n positive real numbers, and let $W = \sum_{i=1}^{n} w_i$. Similarly as for binomial random graphs, we define $\mathcal{G}(\mathbf{w}) = ([n], E)$ to be the probability distribution of graphs (including non-simple graphs) on the set of nodes $[n]$. In the **Chung-Lu model**, each pair of nodes i, j such that $1 \leq i \leq j \leq n$ is independently sampled as an edge (or loop if $i = j$) with probability given by:

$$p_{i,j} = \begin{cases} \frac{w_i w_j}{W}, & i \neq j \\ \frac{(w_i)^2}{2W}, & i = j. \end{cases}$$

Let us mention one technical assumption. In theory, it might happen that $p_{i,j}$ defined above is greater than one. Since the theoretical model allows for parallel edges, if $p_{i,j} > 1$, then it should really be regarded as the expected number of edges between i and j. For example, for $p_{i,j} > 1$ one may introduce a Poisson-distributed number of edges with mean $p_{i,j}$ between each pair of nodes i, j.

However, in practice the maximum weight, $\max_{i \in [n]} w_i$, typically satisfies $(\max_{i \in [n]} w_i)^2 \leq W$, which is the property that is closely related to the problem of the degree sequence being graphic and the structural cut-off mentioned in the previous section. With this assumption we get that $p_{i,j} \leq 1$ and so in practice we rarely face a problem with parallel edges. As a result, in applications we may assume that edges are created with probability $\min\{p_{i,j}, 1\}$ and restrict ourselves to simple graphs, after ignoring loops that are rare but might be present in the original theoretical model, as $p_{i,i} > 0$. Note that, as a result, it slightly biases the expected degree of nodes, especially when w_i is large (that is, it slightly decreases the expectation).

Of course, if $\mathbf{w} = (pn, pn, \ldots, pn)$ and self-loops are ignored, then $\mathcal{G}(\mathbf{w})$ is simply $\mathcal{G}(n, p)$, and so it can be viewed as a generalization of the binomial random graph. One desired property of this random model is that it yields a distribution that preserves the expected degree for each node (exactly for the theoretical non-simple model; or approximately, in practice, for the empirical variant involving simple graphs), namely, for any $i \in [n]$,

$$\mathbb{E}\left[\deg(i)\right] = \sum_{j \in [n]} p_{i,j} = \sum_{j \in [n] \setminus \{i\}} \frac{w_i w_j}{W} + 2 \cdot \frac{(w_i)^2}{2W} = \frac{w_i}{W} \sum_{j \in [n]} w_j = w_i.$$

Not surprisingly, $\mathcal{G}(\mathbf{w})$ is more challenging to analyze than $\mathcal{G}(n,p)$. However, it is also well-studied and many of its interesting properties are known. In order to see another level of complexity, we only mention a few results investigating the appearance of the giant component. For convenience, we assume that $(\max_{i \in [n]} w_i)^2 \leq W$ and use the theoretical model with loops as they do not affect the distribution of connected components.

In this model, it is more natural to call a connected component **giant** if its volume is a positive fraction of the total volume of the graph (see Section 1.8 for a definition). For example, the graph consisting of the binomial random graph $\mathcal{G}(\sqrt{n}, 1/2)$ and $n - \sqrt{n}$ isolated nodes has a giant component a.a.s. but the fraction of nodes in this component vanishes as $n \to \infty$.

Based on experience with binomial random graphs, it is natural to conjecture that there is no giant component if the expected value of the average degree $\langle k \rangle$ is less than 1. However, surprisingly, the answer is negative. For example, if $w_i = 2$ for $i \in [n/2]$ and otherwise $w_i = 0$, then $\mathcal{G}(\mathbf{w})$ is simply a union of $\mathcal{G}(n', 2/n')$ and n' isolated nodes where $n' = n/2$. Since the expected degree in $\mathcal{G}(n', 2/n')$ is $2 > 1$, a.a.s. $\mathcal{G}(n', 2/n')$ has a giant component so the same applies to $\mathcal{G}(\mathbf{w})$. On the other hand, if the expected value of $\langle k \rangle$ is larger than one, then a.a.s. there is a giant component.

So what is going on? Maybe one should look at the second order average degree $\langle k^2 \rangle / \langle k \rangle \geq \langle k \rangle$? Perhaps if the expected value of $\langle k^2 \rangle / \langle k \rangle$ is larger than one, then it is enough to conclude that a.a.s. there is a giant component? Again, the answer is negative. For example, if for some $\epsilon > 0$, $w_i = 1 - \epsilon$ for $i \in [n - \ln n]$ and otherwise $w_i = (1 + \epsilon)\sqrt{n}$, then $\mathcal{G}(\mathbf{w})$ a.a.s. satisfies

$$\langle k \rangle \approx \frac{n - \ln n}{n} (1 - \epsilon) + \frac{\ln n}{n} (1 + \epsilon)\sqrt{n} \sim 1 - \epsilon < 1,$$

and

$$\frac{\langle k^2 \rangle}{\langle k \rangle} \approx \frac{1}{1 - \epsilon} \left(\frac{n - \ln n}{n} (1 - \epsilon)^2 + \frac{\ln n}{n} (1 + \epsilon)^2 n \right) \sim \frac{(1 + \epsilon)^2}{1 - \epsilon} \ln n \gg 1.$$

The graph induced by the set of nodes $[n - \ln n]$ is simply the binomial random graph with expected degree less than one. As a result, a.a.s. all of its components are of size $O(\ln n)$. On the other hand, a.a.s. the max degree is at most, say, $2\sqrt{n}$. So a.a.s. the number of edges in the largest component is at most $O(\sqrt{n} (\ln n)^2) = o(n)$. Since the volume of the original graph is $\Theta(n)$ a.a.s., there is no giant component in $\mathcal{G}(\mathbf{w})$ a.a.s. On the other hand, if $\langle k^2 \rangle / \langle k \rangle$ is less than one, then a.a.s. $\mathcal{G}(\mathbf{w})$ has no giant component.

Hence, the behaviour of **Chung-Lu** model is indeed more complex to analyze, as one may expect from such a general and flexible model. As mentioned above, it can be shown that if $\langle k \rangle \leq \langle k^2 \rangle / \langle k \rangle < 1 - \epsilon$ for some $\epsilon > 0$, then a.a.s. $\mathcal{G}(\mathbf{w})$ has no giant component. On the other hand, if $\langle k^2 \rangle / \langle k \rangle \geq \langle k \rangle > 1 + \epsilon$ for some $\epsilon > 0$, then a.a.s. there is one.

Let us now discuss how one can use the **Chung-Lu** model to generate power-law graphs. In order to generate the expected degree distribution following power-law with degree exponent γ, one may consider

$$w_i = c \cdot (i + i_0 - 1)^{-1/(\gamma-1)}$$

for $i \in [n]$. Here, $c = c(n)$ depends on the minimum (or average) degree $\delta \geq 1$ and $i_0 = i_0(n)$ depends on the maximum degree Δ. It follows that $c = \delta n^{1/(\gamma-1)}$ and $i_0 = n/(\Delta/\delta)^{\gamma-1}$ so

$$w_i = \delta \left(\frac{n}{i - 1 + n/(\Delta/\delta)^{\gamma-1}} \right)^{1/(\gamma-1)}. \tag{2.6}$$

It is slightly technical (and so we omit it) but it is possible to show that the expected number of nodes of degree k is proportional to $\Gamma(k - \gamma + 1)/\Gamma(k + 1) \approx k^{-\gamma}$, where $\Gamma(z) = \int_0^\infty x^{z-1}e^{-x}dx$ is the **gamma function**. A similar property holds for the number of nodes with the expected degree between k and $k + 1$.

In order to illustrate how the model works in practice, we generated $\mathcal{G}(\mathbf{w})$ on $n = 10{,}000$ nodes using the set of weights prescribed by (2.6) with $\gamma = 2.5$, $\delta = 1$, and $\Delta = \sqrt{n} = 100$. The obtained degree distribution is presented in Figure 2.6; in particular, the minimum and the maximum degree of $\mathcal{G}(\mathbf{w})$ were $\delta' = 0$ and $\Delta' = 104$, respectively. Then, in order to estimate the exponent of the power-law degree distribution present in $\mathcal{G}(\mathbf{w})$ we try to find a pair of γ' and tail cut-off level ℓ' that minimize the divergence from the tail of power-law distribution. The divergence measure we use is known in the literature as the **Kolmogorov–Smirnov statistic**. Since we are mostly interested in the properties of degree distribution of large degree nodes, the procedure works as follows. For a given cutoff for small degrees $\ell \in [\max\{\delta, 1\}, \Delta]$, the degree exponent is estimated as follows:

$$\gamma_\ell = 1 + \frac{|\{j : \deg(j) \geq \ell\}|}{\sum_{j:\deg(j)\geq\ell} \ln\left(\frac{\deg(j)}{\ell-1/2}\right)},$$

and the divergence of the distribution from the theoretical distribution can be computed as follows:

$$D_\ell = \max_{k\in[\ell,\Delta]} \left| \frac{|\{j : \deg(j) \geq k\}|}{|\{j : \deg(j) \geq \ell\}|} - \frac{\int_k^\infty x^{-\gamma_\ell}dx}{\int_\ell^\infty x^{-\gamma_\ell}dx} \right|.$$

In the above formula, we used a continuous theoretical distribution to preserve the exposition's consistency; alternatively, one can use a discrete theoretical distribution, an approach that was used in our computational example. The value of γ_ℓ that minimizes D_ℓ (over all $\ell \in [\max\{\delta, 1\}, \Delta]$) is used to estimate γ' of the power-law exponent. Intuitively, we select such a tail cut-off level ℓ' and γ' that minimize the maximum distance between the cumulative distribution function of the power-law distributed tail and the empirical distribution observed in the data.

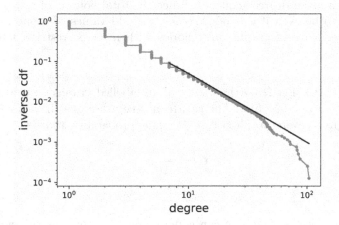

FIGURE 2.6
The inverse cumulative degree distribution (on a log-log scale) of the **Chung-Lu** graph with $n = 10,000$ nodes and $\gamma = 2.5$. The fitted line has slope of -1.70 ($\gamma' = 2.70$), which was obtained with the procedure we described, with $\ell' = 7$.

Finally, let us briefly discuss complexity issues. Similarly to $\mathcal{G}(n, p)$, in order to generate sparse $\mathcal{G}(\mathbf{w})$, one does not need to perform $N = \binom{n}{2} + n$ independent experiments but rather introduce edges one by one and connect two nodes with appropriately selected probabilities. The number of edges to sample is usually well approximated by $\text{Bin}(N, W/(2N))$ or one may simply fix the number of edges to be equal to $\lfloor W/2 \rfloor \approx W/2$, the expected number of edges. In order to generate these random edges, one needs to be able to select nodes randomly but with a non-uniform probability distribution. This step can be done efficiently, as most of the work required for weighted sampling can be done once and then cached, and this part is implemented in Python/Julia and most other computing environments. As we discussed above, this process is only an approximation of a theoretical model if one wants to generate simple graphs. The degrees of nodes with large weights are slightly biased downwards whereas nodes with small weights are biased upwards.

2.6   Random d-regular Graphs

In this section, we consider the uniform probability space of random d-regular graphs with $d \in \mathbb{N} \cup \{0\}$ fixed. Different mathematical tools are required when $d = d(n)$ grows together with n. We will deal with such random graphs in the next section, as they can be viewed as a special case of the family of random graphs with a given degree sequence. Since the total volume of any graph is even, n has to be even if d is odd. Hence, for odd values of d one needs to consider a sequence of graphs on n nodes with $n \rightarrow \infty$ restricted to even numbers.

Fix $d \in \mathbb{N} \cup \{0\}$. Let Ω be the family of all labelled graphs on the set of nodes $[n]$ that are d-regular. The **random d-regular graph**, denoted by $\mathcal{G}_{n,d}$, assigns to every graph $G \in \Omega$ the same probability, that is,

$$\mathbb{P}(G) = \frac{1}{|\Omega|}.$$

Let us first discuss how one can generate these graphs, as this will have some important implications from a theoretical point of view. The cases $d = 0$ and $d = 1$ are trivial. Clearly, 0-regular graphs are simply empty graphs whereas 1-regular graphs are the same as perfect matchings. The random graph $\mathcal{G}_{n,1}$ is obtained by randomly choosing one of the

$$M(n) = \frac{n!}{(n/2)!2^{n/2}} \tag{2.7}$$

partitions of the set of nodes into $n/2$ pairs. Hence, from now on we will assume that d is a fixed integer that is at least 2.

The uniform probability space is very easy to define but is not so easy to work with. After all, nobody is able to generate the family of labelled d-regular graphs on n nodes (even if n is as small as, say, 20) and then select one of them uniformly at random. Hence, instead of working directly in the uniform probability space, we will use the **pairing model** of random regular graphs that is also known as the **configuration model**.

Suppose that dn is even, as in the case of random d-regular graphs, and consider dn **points** partitioned into n labelled **buckets** $v_1, v_2, \ldots, v_n$ of d points each. A **pairing** of these points is a perfect matching into $dn/2$ pairs. Given a random pairing P, we may construct a multigraph $\mathcal{P}_{n,d} = \mathcal{P}(P)$, with loops and parallel edges allowed, as follows. The nodes are the buckets $v_1, v_2, \ldots, v_n$, and a pair $\{x, y\}$ in P corresponds to an edge $v_i v_j$ in $\mathcal{P}_{n,d}$ if x and y are contained in the buckets v_i and v_j, respectively.

There are $M(dn)$ different configurations and a pairing must be selected uniformly at random. This can be done in many different ways, some of which turn out to be very convenient. In particular, the points in the pairs can be chosen sequentially. At any stage of the process, the first point in the next random pair chosen can be selected using any rule whatsoever, as long as the second point in that pair is chosen uniformly at random from the remaining points. For example, when one wants to find nodes at distance ℓ from some given node, a good strategy is to reveal the graph using the **breadth-first search (BFS)** algorithm starting at that node. On the other hand, if the goal is to investigate the size of the connected component containing a given node, then one can utilize the depth-first search algorithm. We will use this approach in our example below.

It is easy to see that the probability that a random pairing corresponds to a given simple graph G is independent of the graph, hence the restriction of $\mathcal{P}_{n,d}$ to simple graphs is precisely $\mathcal{G}_{n,d}$. Indeed, let G_1 and G_2 be any simple graphs on n nodes that are d-regular. Since each simple graph corresponds to precisely $(d!)^n$ pairings, we get immediately that

$$\mathbb{P}\Big(\mathcal{P}_{n,d} = G_1\Big) = \mathbb{P}\Big(\mathcal{P}_{n,d} = G_2\Big) = \frac{(d!)^n}{M(dn)},$$

where $M(dn)$ is defined in (2.7). It follows that a regular graph can be chosen uniformly at random by choosing a pairing uniformly at random and rejecting the result if it has loops or multiple edges. However, non-simple graphs are *not* produced uniformly at random since for each loop the number of corresponding pairings is divided by 2, and for each k-tuple edge it is divided by $k!$.

Moreover, it is known that a random pairing generates a simple graph with probability asymptotic to $e^{-(d^2-1)/4}$ depending on d but not on n. We illustrate this empirically in Figure 2.7. Since in this section we restrict ourselves to graphs with constant degree, this is a desired property as it shows that one needs only a few independent samples to generate a pure instance of $\mathcal{G}_{n,d}$. The expected number of attempts is $e^{(d^2-1)/4}$, which is large for large d but reasonable for relatively small values. Implications of this observation are far more important for theoreticians that aim to prove results that hold a.a.s. in $\mathcal{G}_{n,d}$. Any event holding a.a.s. over the probability space of random pairings $\mathcal{P}_{n,d}$ also holds a.a.s. over the corresponding space $\mathcal{G}_{n,d}$. Indeed, suppose that the property P holds a.a.s. for $\mathcal{P}_{n,d}$. Then, we get that

$$\mathbb{P}(G \in \mathcal{G}_{n,d} \text{ does not have } P)$$
$$= \mathbb{P}(G \in \mathcal{P}_{n,d} \text{ does not have } P \mid G \in \mathcal{P}_{n,d} \text{ is simple})$$
$$= \frac{\mathbb{P}(G \in \mathcal{P}_{n,d} \text{ does not have } P \text{ and is simple})}{\mathbb{P}(G \in \mathcal{P}_{n,d} \text{ is simple})}$$
$$\leq \frac{\mathbb{P}(G \in \mathcal{P}_{n,d} \text{ does not have } P)}{\mathbb{P}(G \in \mathcal{P}_{n,d} \text{ is simple})} \to 0,$$

as $n \to \infty$. For this reason, asymptotic results over random pairings immediately transfer to $\mathcal{G}_{n,d}$. The converse does not hold, as the trivial example of not containing a loop shows.

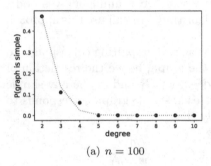
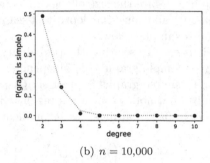

(a) $n = 100$ (b) $n = 10{,}000$

FIGURE 2.7
In order to estimate the probability of $\mathcal{G}_{n,d}$ to be simple, we generated 100 random d-regular graphs with $n = 100$ (a) and $n = 10{,}000$ (b) nodes respectively, and for each value of d such that $2 \leq d \leq 10$. We plot the proportion of simple graphs obtained for each d. The dashed line is the corresponding theoretical prediction.

It is known that a random d-regular graph is a.a.s. connected for any $d \geq 3$. The proof of this fact is involved so we do not present it here. On the other hand, any 2-regular graph consists of a number of disjoint cycles and it is not difficult to see that a random 2-regular graph is a.a.s. disconnected. In fact, we expect the graph to be a relatively large family of cycles. We will show that the total number of cycles Y_n in $\mathcal{G}_{n,2}$ is sharply concentrated near $(1/2) \ln n$. It is not difficult to see this by generating the random graph sequentially using the depth-first search algorithm.

Indeed, let us select any node and select any of the two points p_1, p_2 in it, say, p_1. We will generate a random matching in $\mathcal{P}_{n,2}$, edge by edge, using the depth-first search algorithm. Now, we expose the other point associated with p_1 which is chosen uniformly at random from the remaining points. The probability we create a loop during this very first step is equal to $1/(2n-1)$

as there are $2n - 1$ points that wait to be matched but only one of them creates a loop. If the loop is created, then we move to another node (any node, arbitrarily chosen) and select any point in it. Otherwise (that is if no loop is created), an edge is discovered and the initial point is matched with a point in some other node; we concentrate on the other point and continue the process from there. Regardless of whether a loop is created or not, the probability that we close another cycle in the second step is $1/(2n - 3)$ as there are $2n - 3$ points left and only one of them yields a new cycle. We may repeat this argument until the end of the process so the probability of forming a cycle in step i is exactly $1/(2n - 2i + 1)$.

By linearity of expectation, it follows that the expected number of cycles is equal to

$$\mathbb{E}[Y_n] = \sum_{i=1}^{n} \frac{1}{2n - 2i + 1} = \sum_{j=1}^{2n} \frac{1}{j} - \sum_{j=1}^{n} \frac{1}{2j} = H_{2n} - (1/2)H_n,$$

where H_n is the n-**th harmonic number**. Since $H_n = \ln n + O(1)$, we get that the above is equal to

$$\ln(2n) - (1/2)\ln n + O(1) = \ln n - (1/2)\ln n + O(1) = (1/2)\ln n + O(1).$$

The variance $\mathbb{Var}[Y_n]$ can be calculated in a similar way. Alternatively, the probability that we form a cycle in step i is independent of the history of the process and so some concentration inequalities, such as Chernoff's bound, can be used. In any case, the conclusion is the same: a.a.s. the number of components in $\mathcal{P}_{n,2}$ is asymptotic to $(1/2)\ln n$ so the same property holds for $\mathcal{G}_{n,2}$.

2.7   Random Graphs with a Given Degree Sequence

In this section, we generalize random d-regular graphs to random graphs on the set of nodes $[n]$ with a given desired degree sequence

$$\mathbf{d} = \big(\deg(1), \deg(2), \ldots, \deg(n) \big) = (d_1, d_2, \ldots, d_n). \tag{2.8}$$

Without loss of generality, we may assume that the degree sequence is non-decreasing, that is, $d_1 \leq d_2 \leq \ldots \leq d_n$. Recall that a degree sequence $\mathbf{d}$ is called **graphic** (or **feasible**) if a graph with the degree sequence exists.

Let $\mathbf{d} = (d_1, d_2, \ldots, d_n)$ be any graphic sequence. Let Ω be the family of all labelled graphs on the set of nodes $[n]$ with the degree sequence $\mathbf{d}$. The

random graph with degree sequence d, denoted by $\mathcal{G}_{n,\mathbf{d}}$, assigns to every graph $G \in \Omega$ the same probability, that is,

$$\mathbb{P}(G) = \frac{1}{|\Omega|}.$$

As in the case of random d-regular graphs, we may consider the corresponding pairing model.

Let $\mathbf{d} = (d_1, d_2, \ldots, d_n)$ be any graphic sequence; in particular, $D = \sum_{i=1}^{n} d_i$ is even. Consider D **points** partitioned into n labelled **buckets** $v_1, v_2, \ldots, v_n$; for each $i \in [n]$, bucket v_i consists of d_i points. A **pairing** of these points is a perfect matching into $D/2$ pairs. Given a pairing P, we may construct a multigraph $\mathcal{P}_{n,\mathbf{d}} = \mathcal{P}(P)$, with loops and parallel edges allowed, as follows. The nodes are the buckets $v_1, v_2, \ldots, v_n$, and a pair $\{x, y\}$ in P corresponds to an edge $v_i v_j$ in $\mathcal{P}_{n,\mathbf{d}}$ if x and y are contained in the buckets v_i and v_j, respectively.

Generating $\mathcal{P}_{n,\mathbf{d}}$ can be done efficiently but the resulting graph might not be simple. However, in some applications, the random multigraph may be at least as good as the simple one. If this is the case, then one should simply use $\mathcal{P}_{n,\mathbf{d}}$ which is implemented in Python's iGraph. Moreover, with such assumption in mind, there is no need to make sure that $\mathbf{d}$ is a graphic sequence since, in fact, any sequence with even value of $D = \sum_{i=1}^{n} d_i$ works.

On the other hand, if simple graphs are desired, then one has a number of potential solutions available that we will discuss next. The easiest approach is to remove all potential loops and replace any set of parallel edges with a single edge. This clearly creates a simple random graph but its degree sequence is typically not exactly the given sequence $\mathbf{d}$. Nevertheless, for some applications, this **erased configuration model** may be as useful as more sophisticated solutions.

Suppose now that one insists on generating a random graph with the degree distribution following *exactly* the sequence $\mathbf{d}$. As in the case of random d-regular graphs, the restriction of $\mathcal{P}_{n,\mathbf{d}}$ to simple graphs is precisely $\mathcal{G}_{n,\mathbf{d}}$. Hence, one can apply the resampling algorithm that keeps generating independent copies of $\mathcal{P}_{n,\mathbf{d}}$ until a simple graph is generated. Such graph is precisely $\mathcal{G}_{n,\mathbf{d}}$. This observation is useful from a theoretical point of view but, since the probability that $\mathcal{P}_{n,\mathbf{d}}$ is simple is typically very small, this method is rarely of practical importance. A variant of this method that is implemented in Python's iGraph simply restarts the generation process each time the algorithm gets stuck in a configuration where it is not possible to insert more edges without creating loops or multiple edges. It will succeed eventually, provided that the input degree sequence is graphical. This algorithm is faster than the rejection algorithm mentioned earlier but could still be very slow for some degree distributions. The trade-off is that the outcome of this algorithm is a graph that is not generated uniformly at random from the family of simple graphs with the desired degree distribution **d**. For most applications, this is not a problem.

Another common approach is to start with $\mathcal{P}_{n,\mathbf{d}}$ and adjust it to make it simple using the following **switching algorithm**. In each step, we choose one loop or multiple edges and another random edge, and switch the endpoint of these two edges, thus replacing them with another pair of edges. This algorithm clearly preserves the degree distribution. More importantly, it is known that if both $\sum_{i=1}^{n} d_i$ and $\sum_{i=1}^{n} d_i^2$ are not very far from n, the order of the graph, then this method typically gives a simple graph after a single pass through the bad edges. In fact, in some precise sense (by investigating the total variation distance that is a distance measure for probability distributions), the resulting graph is asymptotically equivalent to $\mathcal{G}_{n,\mathbf{d}}$. One may additionally insist on the resulting graph being connected which also can be obtained by a sequence of edge switchings. Such a Monte-Carlo algorithm is also provided in Python's `iGraph` and uses the original implementation of **Viger**.

Let us now switch gears and discuss the appearance of the giant component in $\mathcal{G}_{n,\mathbf{d}}$. In order to get an intuition about this problem, let us try to provide a heuristic argument that attempts to explain the behaviour of the **breadth-first-search (BFS)** process starting from a given node i. The process discovers the connected component containing node i. We will use the corresponding pairing model $\mathcal{P}_{n,\mathbf{d}}$ and uncover pairs of points, one by one (note that we use $\mathcal{P}_{n,\mathbf{d}}$ not $\mathcal{G}_{n,\mathbf{d}}$ as it is an easier model to work with and our goal is only to present an intuitive argument). As in the case of random d-regular graphs, the first point can be selected using any rule whatsoever as long as the second point is chosen uniformly at random from the remaining points. Hence, in particular, we may explore the graph using the **BFS** procedure.

We initiate the **BFS** algorithm by exposing d_i points associated with node i and putting neighbours of i into the queue Q. Since loops are rare, we expect $(1 + o(1))d_i$ nodes to be present in Q. Assuming that the queue is not empty, the probability that the first node in the queue is node $j \neq i$ is asymptotic to d_j/D, where $D = \sum_{i=1}^{n} d_i$. In the next step of the process, we remove node j from the queue and put all neighbours of j that are not discovered yet into the queue. As triangles are rare, we expect $(1 + o(1))(d_j - 1)$ new nodes joining the queue. Hence, the expected change in the size of the queue is $(1 + o(1))(d_j - 2)$. Note that this can be negative if $d_j = 1$, or asymptotic to zero if $d_j = 2$; in fact, nodes of degree 2 play an important role and we need to pay attention to them. In any case, the expected change in the number of nodes in Q is asymptotic to

$$\frac{\sum_{j \in [n] \setminus \{i\}} d_j(d_j - 2)}{D} \sim \frac{\sum_{j \in [n]} d_j(d_j - 2)}{D} =: \xi(n) = \xi.$$

Thus, it is positive essentially if and only if the sum of the squares of the degrees exceeds twice the sum of the degrees. These calculations only apply to the very first step of the algorithm but the intuition suggests that if only $o(1)$ fraction of nodes is discovered it should be a good prediction for the number of nodes in the queue. Molloy and Reed proved that this intuition captures the behaviour well, subject to certain technical conditions. If $\xi \geq \epsilon$ for some $\epsilon > 0$, then a.a.s. $\mathcal{G}_{n,\mathbf{d}}$ has a giant component. On the other hand, if $\xi \leq -\epsilon$ for some $\epsilon > 0$, then a.a.s. $\mathcal{G}_{n,\mathbf{d}}$ has no giant component.

The imposed technical conditions are quite sophisticated and rarely mentioned in the network science community but they are actually quite important. It might happen that the expected increase may change drastically during the exploration process. Consider, for example, graph on $n = k^2$ nodes for some odd integer k with the degree distribution $d_1 = d_2 = \ldots = d_{n-1} = 1$ and $d_n = 2k$. Clearly, the only graph that occurs is the disjoint union of a star $K_{1,2k}$ and $(n-2k-1)/2$ isolated edges, hence (deterministically) the giant component has order $2k+1 = o(n)$. On the other hand, $D = (n-1)+2k \sim n = k^2$ and

$$\xi = \frac{-(n-1) + (2k)(2k-2)}{D} \sim \frac{3k^2}{k^2} = 3,$$

so the Molloy-Reed approach would suggest that the giant component exists a.a.s. This shows that, indeed, the technical condition cannot be neglected and should be taken into account.

In order to state the criterion for the existence of the giant component, we need to introduce the following definitions:

$$j_{\mathbf{d}} = \min \left(\left\{ j : j \in [n] \text{ and } \sum_{i=1}^{j} d_i(d_i - 2) > 0 \right\} \cup \{n\} \right),$$

$$R_{\mathbf{d}} = \sum_{i=j_{\mathbf{d}}}^{n} d_i, \quad \text{and}$$

$$M_{\mathbf{d}} = \sum_{i \in [n]:d_i \neq 2} d_i.$$

The existence of a giant component essentially depends on whether $R_{\mathbf{d}}$ is of the same order as $M_{\mathbf{d}}$ or not. However, things break down if almost all nodes have degree 2 so we assume that $M_{\mathbf{d}}$ tends to infinity together with n. Now, we are ready to present the state-of-the-art result. If $R_{\mathbf{d}} = o(M_{\mathbf{d}})$, then a.a.s. $\mathcal{G}_{n,\mathbf{d}}$ has no giant component. On the other hand, if $R_{\mathbf{d}} = \Omega(M_{\mathbf{d}})$, then a.a.s. $\mathcal{G}_{n,\mathbf{d}}$ has a giant component.

Coming back to our example with the star and isolated edges, note that for that particular degree distribution we have $j_{\mathbf{d}} = n = k^2$, $R_{\mathbf{d}} = 2k$, and $M_{\mathbf{d}} = D \sim k^2$. It follows that $R_{\mathbf{d}} = o(M_{\mathbf{d}})$, and the result implies that a.a.s. there is no giant component.

2.8 Experiments

We revisit one of the graphs we experimented with in Chapter 1, namely, the GitHub developers graph. We consider the giant component of the machine learning (ml) developers subgraph, which consists of 7,083 nodes and 19,491 edges. Based on this base graph, we generate a few random graphs using different models we introduced in this chapter, namely:

- the **binomial random graph** model $\mathcal{G}(n, m)$ which only preserves the average degree (Section 2.3),

- the **Chung-Lu** model $\mathcal{G}(\mathbf{w})$ which preserves the expected degree for each node (Section 2.5), and

- the **configuration model** in which the degree distribution is preserved (Section 2.7).

For the configuration model, we consider the default implementation in iGraph which may introduce multi-edges, loops, or create a disconnected graph (our $\mathcal{P}_{n,\mathbf{d}}$) as well as the version implemented by **Viger** where a connected simple graph is generated. We summarize some basic statistics in Table 2.8. As expected, advanced models preserve more statistics of the original graph. We concentrate on that are preserved increases. In particular, the configuration model of **Viger** almost perfectly matches the basic statistics of the original graph with the only difference being the diameter and the clustering coefficients.

In Figure 2.9, we compare the shortest path lengths distribution of the base graph with the 4 random graphs we experimented with earlier. We generated those distributions by computing the shortest path length between every pair of nodes. We see a reasonably high similarity for all graphs, with the binomial random graph having slightly longer path lengths due to the absence of high degree (hub) nodes in that model.

2.9 Practitioner's Corner

Random graph models are of theoretical interest and this is an extremely active research field that attracts not only mathematicians but also physicists

TABLE 2.8

Comparison of a few descriptive statistics for the base graph (a subgraph of the GitHub graph) with 4 random graph models.

Graph	Base	Binomial	Chung-Lu	Config.	Config.(V)
nodes	7,083	7,083	7,083	7,083	7,083
edges	19,491	19,491	19,491	19,491	19,491
$\delta = d_{min}$	1	0	0	1	1
d_{mean}	5.504	5.504	5.504	5.504	5.504
d_{median}	2	5	3	2	2
$\Delta = d_{max}$	482	18	406	482	482
diameter	13	10	11	10	11
components	1	25	1,063	70	1
largest	7,083	7,058	5,992	6,940	7,083
isolates	0	23	1,035	0	0
C_{glob}	0.0338	0.0007	0.0198	0.0185	0.0171
C_{loc}	0.1412	0.0006	0.0258	0.0242	0.0319

and computer scientists. They are also often used in practice to generate synthetic graphs that are similar to real-world graphs and so can be used to test various hypotheses or algorithms. This is particularly useful when real-world graphs are scarce.

Depending on the type of application, it may be enough to generate graphs with a given number of nodes and edges (and so also a given average degree), in which case the **binomial random** graph model may be used. If one wants more realistic graphs, for example, with a mix of high and low degree nodes, then models such as **Chung-Lu** and the **configuration model** may be used. **Chung-Lu** may yield isolated nodes, and only the expected degrees are preserved. On the other hand, with the **configuration model**, one may preserve the degree sequence exactly. Moreover, if one uses the **Viger**'s generator for this model to produce a connected graph, then not only will the degree distribution be preserved but the distribution of orders of connected components will be maintained as well. Indeed, in order to generate such a graph, one may independently generate connected graphs that match the degree distribution of all connected components of the original graph and then simply take their union.

Of course, the models we introduced in this chapter only match some basic properties of the base graph. Depending on the application at hand, one may introduce more sophisticated models that preserve more properties and so mimic the structure of the base graph to a higher degree. We introduce examples of such models in the following chapters.

Finally, let us make a comment about the degree distributions observed in real-world networks. It is quite common that these networks have degree distributions following the power-law. That is why in Section 2.5 we discussed

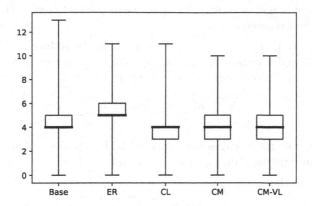

FIGURE 2.9
Comparison of the distribution of shortest path lengths for the giant component of the base graph with the graphs generated from 4 random graph models.

how one can model such distributions, given the degree exponent γ and the average degree $\langle k \rangle$. However, not all real-world networks follow such distributions. Indeed, let us apply the same procedure of finding the exponent of power-law distributions for other graphs introduced in Chapter 1. For the GitHub (ml) graph studied here, we found the exponent of $\gamma = 2.62$ and for the GitHub (web) $\gamma = 2.51$. Such values agree with power-law distributions observed in social networks. However, for the Grid graph, we found $\gamma = 19$ as the estimate for the degree exponent with a best fit obtained for only three nodes, a clear indication that the degree distribution for this graph does *not* follow a power-law.

2.10 Problems

In this section, we present a collection of potential practical problems for the reader to attempt.

1. Implement an algorithm that checks whether a given degree sequence is graphic.

2. Generate a degree sequence following the power law distribution (try different degree exponents, maximum, minimum degrees, etc.). Compare the expected number of edges in the **Chung-Lu** model

with parallel edges and loops and the corresponding expected value for the variant in which we replace each parallel edge with a single edge and remove loops.

Recall that for $p_{i,j} > 1$ one typically introduces a Poisson-distributed number of edges with mean $p_{i,j}$ between each pair of nodes i, j. So one may simply compute

$$\sum_{1 \le i < j \le n} \max\{p_{i,j} - 1, 0\} + \sum_{1 \le i \le n} p_{i,i}$$

to see how many edges are expected to be removed. Alternatively, one may actually do an experiment and check it.

3. Suppose that
$$p = p(n) = \frac{\ln n + \ln \ln n + c}{2n}$$
    for some constant $c \in \mathbb{R}$. The expected number of isolated edges in $\mathcal{G}(n, p)$ (an edge uv is isolated if u is the only neighbour of v and vice versa) is asymptotically equal to $e^{-c}/4$.

    Compare the above theoretical prediction for the expected number of isolated edges with empirical results based on 1,000 independent runs for small graphs on $n = 100$ nodes and larger graphs on $n = 10,000$ nodes. Make a figure similar to Figure 2.2 with, for example, $c \in [-3, 3]$.

4. We showed that the expected number of cycles in $\mathcal{P}_{n,2}$ is asymptotic to $(1/2) \ln n$. Compare this theoretical prediction with empirical results based on 1,000 independent runs for small graphs on $n = 100$ nodes and larger graphs on $n = 10,000$ nodes.

5. We showed that a.a.s. $\mathcal{G}_{n,3}$ is connected. Compare this theoretical prediction with empirical results based on 1,000 independent runs for small graphs on $n = 100$ nodes and larger graphs on $n = 10,000$ nodes.

6. Generate two independent copies of $\mathcal{G}(n, p)$ model with $n = 10,000$ and $p = 1/4$, graphs G_1 and G_2. Then take a union of these two graphs, $G_1 \cup G_2$ (edge uv is present in $G_1 \cup G_2$ if it is present in at least one of the two involved graphs). Check the density of $G_1 \cup G_2$, that is, the ratio between the number of edges in $G_1 \cup G_2$ and $\binom{n}{2}$. Is it close to $p + p = 1/2$? If not, can you explain why?

7. The **Watts–Strogatz** model generates networks that have a small diameter and a large clustering coefficient. It starts with a grid network where all nodes have the same number of "close" neighbours. Each node is then independently "rewired" with probability $p \in [0, 1]$. If an edge is rewired, one of its endpoints is preserved and the other one is replaced by a node selected uniformly at random from the set of all nodes. This typically creates a "long" edge.

This model is implemented in `igraph` (`Watts_Strogatz`). Perform an experiment with $dim = 2$ (dimension of the lattice), $size = 100$ (the size of the lattice along each dimension), $nei = 8$ (the average degree). Plot the global clustering coefficient and the average local clustering coefficient as a function of $p \in (0,1)$. You may want to do a few repetitions for each value of p to smooth the graph.

8. The **Preferential Attachment** model uses the "rich get richer" principle to generate graphs with a power-law degree distribution. This model is implemented in `igraph` (`Barabasi`). Perform an experiment with $n = 10,000$ (the number of nodes) and $m = 5$ (the number of directed edges created in each step). Plot the inverse cumulative in-degree distribution. Use the Kolmogorov–Smirnov test to estimate the degree exponent.

2.11   Recommended Supplementary Reading

The field of random graphs is an active area of research with thousands of papers written on the topic. Here is a list of a few books:

- B. Bollobás, *Random Graphs*, Cambridge University Press, 2nd Edition, 2011.

- S. Janson, T. Łuczak, A. Ruciński, *Random Graphs*, Wiley-Interscience, 2000.

- A. Frieze, M. Karoński, *Introduction to Random Graphs*, Cambridge University Press, 2015.

- F. Chung, L. Lu, *Complex Graphs and Networks*, American Mathematical Society, 2006.

- R. van der Hofstad, *Random Graphs and Complex Networks. Volume 1*, Cambridge Series in Statistical and Probabilistic Mathematics, 2017.

In order to fit the power-law curve, we used the `plfit` package[1] based on the following paper:

- A. Clauset, C.R. Shalizi, and M.E.J. Newman, *Power-law distributions in empirical data*, SIAM Review 51(4), 661–703 (2009).

---

[1]`pypi.org/project/plfit/`

3

Centrality Measures

3.1 Introduction

In this chapter, our goal is to identify the most important nodes within a graph. This is an important task from the application's perspective. Potential applications include finding the most influential users in a social network, the busiest intersections in a large city such as Toronto, key infrastructure nodes in the energy network, and ranking web pages returned by a search engine.

There are clearly many ways to measure how central a given node is by comparing it with other nodes that are present in the graph. We will propose a few natural ways to do it and perform a number of experiments. Each node will, depending on the chosen approach, receive a "score" $c(v)$. It is sometimes enough to rank the nodes with respect to their scores. However, in order to be able to compare various measures, the scores should be normalized so that they are real numbers between zero and one. Additionally, for some centrality measures, it might be convenient to assume that $\sum_{v \in V} c(v) = 1$. We sometimes explicitly do it but this additional property can always be enforced, if needed. We will try to follow the normalization implemented in the `igraph` package of Python.

Having flexible scores assigned to all nodes is useful and informative but, in practice, we often face a problem of identifying the most central node or, say, a set of top 10 most central nodes. For example, we might need to find the top researchers within the collaboration network to be invited to the conference we organize, or our goal might be to identify top 10 most influential bloggers and send our new product to them for their on-line review. As a result, we often need to rank all the nodes based on a given centrality measure.

This chapter is structured as follows. We first discuss centrality measures that use algebraic properties of the adjacency matrix associated with the graph (Section 3.2). The next family of measures we investigate are based on paths between pairs of nodes (Section 3.3). Then, since there are many different centrality measures to choose from, we provide a few ways to compare them (Section 3.4). It is often the case that identifying the most central nodes is difficult. On the other hand, there are many nodes that are clearly not central. We show how to prune them using the notion of k-cores (Section 3.5). After that we discuss how to identify a central group of nodes and how to measure

DOI: 10.1201/9781003218869-3

the distribution of centrality within the graph (Section 3.6). As usual, we finish the chapter with experiments (Section 3.7) and provide some tips for practitioners (Section 3.8).

3.2   Matrix Based Measures

Degree Centrality

The first centrality measure is very simple and tries to capture the fact that a node with many neighbours can be seen as more central since it simply has a direct access to more nodes. Depending on the application, it might indicate better access to information or resources, being more popular, influential, or prestigious, etc. We start with a definition for undirected graphs.

Let $G = (V, E)$ be any undirected graph on n nodes. The **degree centrality** of a node $v \in V$ is defined as follows:

$$c_d(v) = \frac{\deg(v)}{n - 1} = \frac{1}{n - 1} \sum_{u \in V} a(v, u).$$

Since the degree of each node is always a non-negative real number (or, in fact, an integer in the case of unweighted graphs) that is at most $n - 1$, the degree centrality is properly normalized such that it is always between zero and one.

For sparse and unweighted graphs, instead of using the degree of v as a measure of its centrality, one can use the number of nodes at distance at most k for some fixed parameter $k \in \mathbb{N}$. In other words, $c(v) = |N_{\leq k}(v)|/(n - 1)$, where $N_{\leq k}(v) = \{u \in V : \text{dist}(v, u) \leq k\}$ is the set of nodes at distance at most k from v.

Generalizing this centrality measure to directed graphs is straightforward. Depending on the application in mind, one might care more about in-, out-, or total degree.

Let $D = (V, E)$ be any directed graph on n nodes. The **degree central-ities** of a node $v \in V$ are defined as follows:

$$c_d^{in}(v) = \frac{1}{n-1} \sum_{u \in N^{in}(v)} a(u, v) = \frac{1}{n-1} \sum_{u \in V} a(u, v)$$

$$c_d^{out}(v) = \frac{1}{n-1} \sum_{u \in N^{out}(v)} a(v, u) = \frac{1}{n-1} \sum_{u \in V} a(v, u)$$

$$c_d^{tot}(v) = \frac{1}{2} \left(c_d^{in}(v) + c_d^{out}(v) \right).$$

Eigenvector Centrality

Let us now consider the following question. Suppose that one node has rela-tively high degree but its neighbours are not significant within the network. On the other hand, some other node has relatively small degree but its neigh-bours are highly connected and influential. Which of the two should be ranked higher? In the extreme case, suppose that Robert had only one friend but his friend happened to be Isaac Newton, which made Robert quite important and influential. Indeed, connections to high-scoring nodes should contribute more to the score of the node in question than a similar number of connections to low-scoring nodes.

As before, let us start with undirected graphs. Keeping our previous dis-cussion in mind, the goal of the next centrality measure is to assign a score $c(v)$ to a node v that is proportional to the sum of the centralities of its neighbours, that is,

$$c(v) = \frac{1}{\lambda} \sum_{u \in N(v)} c(u) \tag{3.1}$$

for some constant $\lambda \in \mathbb{R}_+$. Equation (3.1) is equivalent to $\lambda c(v) = \sum_{u \in V} a(v, u) c(u)$, where $\mathbf{A} = (a(u, v))_{u,v \in V}$ is the adjacency matrix of G. Moreover, it can be written in matrix form as follows: $\lambda \mathbf{c} = \mathbf{A} \mathbf{c}$ or

$$(\mathbf{A} - \lambda \mathbf{I})\mathbf{c} = 0, \tag{3.2}$$

where $\mathbf{c} = (c(v_1), \dots, c(v_n))$ is the unknown vector of centrality measures. This is a familiar problem from linear algebra: if constant λ and vector $\mathbf{c}$ exist, then we call them an **eigenvalue** and, respectively, an **eigenvector** of matrix $\mathbf{A}$.

Note that the system of equations in (3.2) is a **homogeneous** system of linear equations. Hence, in particular, $\mathbf{c} = \mathbf{0} = (0, \dots, 0)$ is a trivial solution that we are clearly not interested in. Indeed, our goal is to find a positive eigenvalue $\lambda \in \mathbb{R}_+$ and the associated eigenvector $\mathbf{c}$ with all entries being positive real numbers. In order to get a non-trivial solution, we need to select λ such that

$$p(\lambda) := \det(\mathbf{A} - \lambda \mathbf{I}) = 0;$$

$p(\lambda)$ is called the **characteristic polynomial** of matrix $\mathbf{A}$. In other words, the eigenvalues of $\mathbf{A}$ are simply the roots of the characteristic polynomial $p(\lambda)$. Since $p(\lambda)$ is a polynomial of degree n, the **Fundamental Theorem of Algebra** implies that there are exactly n roots that are complex numbers (counting multiplicities appropriately). Moreover, since $\mathbf{A}$ is real and symmetric (recall that G is undirected), all eigenvalues are real. We say that an eigenvalue is a **leading eigenvalue** if it is positive and greater than or equal to (in absolute value) all other eigenvalues.

We have many choices for λ but what about the additional requirement that all the entries in the eigenvector be non-negative? Fortunately, the famous **Perron–Frobenius Theorem** implies that the leading eigenvalue is unique and is the only eigenvalue that yields the desired centrality measure, provided that graph G is connected (which is equivalent to matrix $\mathbf{A}$ being **irreducible**—we omit the definition as it is not important for further discussion).

Finally, let us mention that the eigenvectors are only defined up to a common multiplicative factor, so only the ratios of the centralities of the nodes are well defined. In order to define an absolute score and make sure all scores are in the interval $[0, 1]$, one needs to normalize them so that the largest score is equal to one. This leads us to the following definition.

Let $G = (V, E)$ be any connected undirected graph on n nodes. Let λ be the leading eigenvalue of the adjacency matrix $\mathbf{A}$ of graph G, and let $\mathbf{c}$ be the associated non-negative eigenvector; in particular, $\lambda \mathbf{c} = \mathbf{A}\mathbf{c}$. The **eigenvector centrality** of a node $v \in V$ is defined as follows:

$$c_e(v) = \frac{c(v)}{\max_{u \in V} c(u)}.$$

Let us now move to directed graphs and try to adjust the above definition for them. It can be done but it creates a number of issues that we will discuss next. Because of that, the eigenvector centrality is often used for undirected graphs and less often for directed ones. Nevertheless, it will be useful to identify and discuss the issues in order to better understand the next centrality measures and the reasons that they are defined the way that they are.

Let us first note that a directed graph has, in general, an asymmetric adjacency matrix. As a result, it has two sets of eigenvectors: the left eigenvectors and the right eigenvectors. A **right eigenvector** is a column vector satisfying $\mathbf{A}\mathbf{c} = \lambda \mathbf{c}$ for some λ. On the other hand, a **left eigenvector** is a row vector satisfying $\mathbf{c}\mathbf{A} = \lambda \mathbf{c}$ for some λ. Note that a left eigenvector of $\mathbf{A}$ is the same as the transpose of a right eigenvector of $\mathbf{A}^T$, with the same eigenvalue. Moreover, since the characteristic polynomials of $\mathbf{A}$ and $\mathbf{A}^T$ are the same, the eigenvalues of the left eigenvectors of $\mathbf{A}$ are the same as the eigenvalues of the right eigenvectors of $\mathbf{A}^T$.

Coming back to our discussion, for directed graphs we now have two leading eigenvectors associated with the leading eigenvalue to choose from, one right and one left. Which one should we use to define the centrality measure? It might depend on the application at hand but typically the reason for a node to be central is that many other central nodes point towards it rather than it pointing to many central nodes. For example, Cristiano Ronaldo (216M+ followers) and Ariana Grande (183M+ followers) are considered to be the most popular users on Instagram (May 2020) because of a large number of followers, regardless of whether they follow many other users or not. Similar arguments can be applied to many other directed networks. Hence, the most natural choice is to use the right leading eigenvector.

The second issue is with nodes of in-degree zero. Clearly, such nodes have centralities equal to zero. This, arguably, is acceptable but, unfortunately, it propagates throughout the graph. Such nodes of in-degree zero could have many out-neighbours which, in turn, should increase their own centralities. However, only nodes that are in a strongly connected components consisting of at least two nodes, or nodes in the out-component of one of such strongly connected components, can have non-zero centrality. This motivates the next definition and also solves the problem of uniqueness of a leading eigenvalue as it implies that the adjacency matrix is irreducible (again, we omit details as they are not important for our purpose). However, as already mentioned, for directed graphs it is recommended to use either the Katz centrality or PageRank centrality measures that we will discuss next. Indeed, many directed networks are *not* strongly connected. For example, the web consists of the strongly connected component (GSCC), the IN set of nodes that can reach GSCC but which cannot be reached from GSCC, the OUT set of nodes that can be reached from GSCC but which cannot reach GSCC and "tendrils" that are reachable from some nodes in the IN set or that can reach some nodes of the OUT set, without passing through GSCC. Based on the very first study done in 2000, these four sets consist of roughly 25% nodes of the network. However, it is worth pointing out that these proportions strongly depend on the crawling process and other values have been reported since then. Finally, "tubes" consist of nodes that pass from some part of the IN set to some part of the OUT set without touching GSCC.

Let $D = (V, E)$ be any strongly connected, directed graph on n nodes. Let λ be the leading eigenvalue of the adjacency matrix $\mathbf{A}$ of graph D, and let $\mathbf{c}^{in}$ and $\mathbf{c}^{out}$ be the non-negative eigenvectors associated with the leading eigenvalue of matrix $\mathbf{A}$; in particular, $\lambda\mathbf{c}^{in} = \mathbf{A}\mathbf{c}^{in}$ and $\lambda\mathbf{c}^{out} = \mathbf{A}^T\mathbf{c}^{out}$.

The **eigenvector centralities** of a node $v \in V$ are defined as follows:

$$c_e^{in}(v) = \frac{c^{in}(v)}{\max_{u \in V} c^{in}(u)} \qquad \text{and} \qquad c_e^{out}(v) = \frac{c^{out}(v)}{\max_{u \in V} c^{out}(u)}.$$

Katz Centrality

In order to solve the issue we discussed above with nodes of in-degree zero, we may simply assign a small amount of centrality to each node "for free," regardless of the structure of the graph and position of this node within the graph. That is,

$$c(v) = \alpha \sum_{u \in V} c(u)a(u,v) + \beta \qquad (3.3)$$

for some positive constants α and β. As a result, each node v will have $c(v) \geq \beta$ and so any node that is pointed to by many other nodes will have a high score, even if it does not belong to a strongly connected component or out-component. Since we do not care about the absolute value of $c(v)$ (it will be eventually normalized anyway), we may assume that $\beta = 1$. Equation (3.3) can be re-written using matrix form as follows:

$$\mathbf{c} = (\mathbf{I} - \alpha \mathbf{A}^T)^{-1} \mathbf{1},$$

where $\mathbf{1} = (1, \ldots, 1)$ is the uniform vector.

This centrality measure is a function of the parameter α and so it is also often called α-**centrality**. This parameter guides the relative importance of the graph structure. Indeed, in the extreme case when $\alpha = 0$ it produces equal scores for all nodes, regardless of the structure of the graph. On the other hand, large values of α imply that the structure is crucial. There is a technical upper bound for α to make sure that the matrix $\mathbf{I} - \alpha \mathbf{A}^T$ is invertible, namely, $\alpha < 1/|\lambda|$, where λ is the leading eigenvalue of $\mathbf{A}$. However, the connection to the attenuation factor, which we will discuss soon, justifies the usual convention to additionally require that $\alpha < 1$.

The Katz centrality measure can be applied to both undirected and directed graphs.

Let $D = (V, E)$ be any directed graph on n nodes. Alternatively, let $G = (V, E)$ be any graph on n nodes. Fix any α such that $0 < \alpha < \min\{1, 1/|\lambda|\}$, where λ is the leading eigenvalue of $\mathbf{A}$. The **Katz centrality** of nodes in the graph is defined as follows:

$$\mathbf{c}_\alpha = (\mathbf{I} - \alpha \mathbf{A}^T)^{-1} \mathbf{1}.$$

As promised earlier, we will now make a connection to the attenuation factor which gives us another natural interpretation of the Katz centrality measure. This point of view will, in addition, provide a justification for not rescaling Katz centrality as we do with the other centrality measures we have discussed so far. We will show that

$$c_\alpha(v) = \sum_{k=0}^{\infty} \sum_{u \in V} \alpha^k \mathbf{A}^k(u, v). \tag{3.4}$$

However, before we do so, let us interpret it. First, note that for unweighted graphs, $\mathbf{A}^k$, the matrix product of k copies of $\mathbf{A}$, has an interesting and important interpretation: the element (u, v) of $\mathbf{A}^k$ is equal to the number of (directed or undirected) walks of length k from node u to node v. It follows that the Katz centrality computes the relative influence of a node within a network by considering walks of any length. Connections made with distant neighbours are, however, penalized by the **attenuation factor** α applied to each edge of the walk. As a result, indeed, a natural assumption is that $\alpha < 1$ since we want to penalize longer walks, not make them more attractive. In other words, suppose that each node of a graph has a token (think of it as a piece of information). In each step of the process, each token occupying some node of the graph is replicated and sent to all neighbouring nodes with probability α; otherwise, it is simply destroyed. This process is repeated recursively. The Katz centrality of a node is then equal to the expected number of all visits to this node by all tokens in any step of this process (including the very first deterministic visit when an initial placement of tokens is made).

Now, let us show that equation (3.4) holds. In order to see this, observe that it can be rewritten as follows:

$$\mathbf{c}_\alpha^T = \sum_{k=0}^{\infty} \alpha^k \mathbf{1}^T \mathbf{A}^k = \mathbf{1}^T \sum_{k=0}^{\infty} (\alpha \mathbf{A})^k = \mathbf{1}^T (\mathbf{I} - \alpha \mathbf{A})^{-1}$$

and, after transposition, we get precisely the definition of the Katz centrality. The key point to highlight here is that

$$(I - \alpha \mathbf{A}) \sum_{k=0}^{\infty} (\alpha \mathbf{A})^k = \sum_{k=0}^{\infty} (\alpha \mathbf{A})^k - \sum_{k=1}^{\infty} (\alpha \mathbf{A})^k = \mathbf{I}$$

which clearly holds if the above series is convergent. This property is ensured since parameter α is assumed to be less than the inverse of the absolute value of the leading eigenvalue of $\mathbf{A}$.

PageRank Centrality

One potentially undesirable property of the Katz centrality measure is that a high-centrality node pointing to many other nodes immediately gives them a high score. One could argue that this is not always an appropriate feature.

The fact that, for example, `amazon.com` webpage puts a hyperlink to my web page does not mean much as they have an enormous number of outgoing links. On the other hand, if Cristiano Ronaldo follows me on Instagram, then this fact alone should affect my own centrality as he currently (May 2020) follows only 447 users.

Indeed, arguably, it makes more sense for a node in a graph to pass only a small fraction of its own centrality score to each of its neighbours, inversely proportional to its out-degree. That is,

$$c(v) = \alpha \sum_{u \in V} c(u)\hat{a}(u,v) + \beta, \tag{3.5}$$

for some positive constants α, β, and matrix $\hat{\mathbf{A}} = (\hat{a}(u,v))_{u,v \in V}$ defined as follows:

$$\hat{a}(u,v) = \begin{cases} a(u,v)/\deg^{out}(u) & \text{if } \deg^{out}(u) > 0 \\ 1/n & \text{otherwise.} \end{cases} \tag{3.6}$$

The definition of $\hat{a}$ implies that if some node has out-degree zero, then it transfers its weights equally among all nodes (including itself). This way, in particular, we make sure that we avoid dividing by zero.

Arguing as before, we may assume that $\beta = 1$ and so, after rearranging equality (3.5) and using matrix form, we get that:

$$\mathbf{c} = \left(\mathbf{I} - \alpha\hat{\mathbf{A}}^T\right)^{-1} \mathbf{1}.$$

Note that $\|\mathbf{c}\| = n/(1-\alpha)$. As with the Katz centrality measure, there is a positive parameter α which has to be tuned. Arguing as before, we get that the value of α has to be less than the inverse of the largest eigenvalue of $\hat{\mathbf{A}}$ which is equal to one as this is a stochastic matrix, that is, a real square matrix, with each row summing to one. For example, the Google search engine used $\alpha = 0.85$ for their initial algorithm but it is not clear if there is any rigorous theory behind this choice.

Let $D = (V, E)$ be any directed graph on n nodes. Alternatively, let $G = (V, E)$ be any graph on n nodes. Fix any $0 < \alpha < 1$. The **PageRank centrality** of nodes in the graph is defined as follows:

$$\mathbf{c}_p = \frac{1-\alpha}{n} \left(\mathbf{I} - \alpha\hat{\mathbf{A}}^T\right)^{-1} \mathbf{1},$$

where matrix $\hat{\mathbf{A}}$ is defined in (3.6).

There is another interpretation and theory behind PageRank. Suppose that an imaginary web surfer is randomly clicking on hyperlinks but eventually gets bored by the current website and surfs to a new random site. At any step, the probability that the person continues browsing is a **damping factor** α; that is, with probability $1 - \alpha$ the person selects a new site uniformly at random from the set of all sites. This process can be understood using a theory of Markov chains in which the states are web pages and the transitions are the hyperlinks between pages, all of which are equally probable. The PageRank centrality measure of a web page reflects the probability that the random surfer will visit that page at some time in the future, after sufficiently many steps. Formally, this is defined as the stationary distribution of the corresponding Markov chain. Informally, the websites are ranked by how many times they were visited during this process. The intuition is that websites are visited more often if they are linked by many other sites, which should be a good measure of how important a website is.

Hubs and Authorities

As argued above, the reason for a node to be considered central is usually that many other important nodes send a link towards it. However, there might be some applications where this is not the case. Consider, for example, the citation graph where nodes correspond to scientific papers and directed edges correspond to references between them. Clearly, papers that are often cited should be considered central but there are other papers (for example, surveys or reviews) that are not cited too often but they by themselves cite a large number of other important papers. Such papers might contain relatively little new content on the subject but they are certainly great resources to learn where the desired information can be found. As a result, they are useful and important nodes within the collaboration graph.

The scenario discussed above suggests that there are really two types of important nodes: authorities and hubs. **Authorities** are nodes that contain important information on a topic of interests. On the other hand, **hubs** are nodes that reveal where the best authorities are. Of course, some nodes could be good authorities as well as good hubs. Moreover, let us point out that this distinction can only be made for directed graphs and that this concept does not apply to undirected graphs.

The **HITS** algorithm (**Hyperlink-Induced Topic Search**) assigns to each node $v \in V$ of a directed graph $D = (V, E)$ two different centrality measures: the **authority centrality** $x(v)$ and the **hub centrality** $y(v)$. The authority centrality of a node v should be affected by the hub centralities of the nodes that point to it, that is,

$$x(v) = \alpha \sum_{u \in V} y(u)a(u,v)$$

for some positive constant α. Similarly, the hub centrality should be a function of the authority centralities of the nodes that it points to, that is,

$$y(v) = \beta \sum_{u \in V} a(v,u)x(u)$$

for some positive constant β. These equations can be rewritten in matrix form as follows:

$$\mathbf{x} = \alpha \mathbf{A}^T \mathbf{y}$$
$$\mathbf{y} = \beta \mathbf{A} \mathbf{x}. \qquad (3.7)$$

After combining the two we get

$$\lambda \mathbf{x} = \mathbf{A}^T \mathbf{A} \mathbf{x} \qquad (3.8)$$
$$\lambda \mathbf{y} = \mathbf{A} \mathbf{A}^T \mathbf{y}, \qquad (3.9)$$

where $\lambda = 1/(\alpha\beta)$. It follows that $\mathbf{x}$ and $\mathbf{y}$ are eigenvectors of $\mathbf{A}^T \mathbf{A}$ and, respectively, $\mathbf{A}\mathbf{A}^T$.

Let us first point out that, in general, these equations could have multiple solutions. For instance, consider the following adjacency matrix

$$\mathbf{A} = \begin{bmatrix} 0 & 0.5 \\ 1 & 0 \end{bmatrix}$$

for which we have

$$\mathbf{A}^T \mathbf{A} = \begin{bmatrix} 1 & 0 \\ 0 & 0.25 \end{bmatrix} \quad \text{and} \quad \mathbf{A}\mathbf{A}^T = \begin{bmatrix} 0.25 & 0 \\ 0 & 1 \end{bmatrix}.$$

Both $\mathbf{A}^T \mathbf{A}$ and $\mathbf{A}\mathbf{A}^T$ have the same eigenvalues, namely, $\lambda_1 = 0.25$ and $\lambda_2 = 1$. (We will show below that this is not a coincidence.) For $\lambda_1 = 0.25$, after normalizing the vectors, we get the following solution: $\mathbf{x} = (0,1)$ and $\mathbf{y} = (1,0)$. On the other hand, for $\lambda_2 = 1$, after normalizing the vectors we get: $\mathbf{x} = (1,0)$ and $\mathbf{y} = (0,1)$. Both solutions yield non-negative scores and so could be used for our purpose.

Another issue is that $\mathbf{A}^T \mathbf{A}$ might not have a unique leading eigenvalue; for example, consider a directed graph on two nodes, u and v, with $a(u,v) = a(v,u) = 1$. However, this condition is usually met in practice so we will assume it in what follows; when the assumption does not hold, then there is no unique solution in non-negative normalized vectors to the given set of the equations. For instance, in our example of a directed graph with two nodes linked by an edge, any vectors such that $\mathbf{x} = \mathbf{y} = (\sin(t), \cos(t))$, where $t \in [0, \pi/2]$, are non-negative and normalized solutions to the given set of equations.

As in the case of earlier centrality measures, we will consider the non-negative eigenvector that corresponds to the associated unique leading eigenvalue. (As $\mathbf{A}^T \mathbf{A}$ is symmetric, this eigenvector can be made non-negative). However, for this approach to work, we need $\mathbf{A}\mathbf{A}^T$ and $\mathbf{A}^T \mathbf{A}$ to have exactly the same leading eigenvalue λ. Fortunately, it is the case and, in fact, all eigenvalues are the same. Indeed, suppose that λ is an eigenvalue of $\mathbf{A}\mathbf{A}^T$, that is, $\mathbf{A}\mathbf{A}^T \mathbf{y} = \lambda \mathbf{y}$. Then, after multiplying both sides by $\mathbf{A}^T$, we get that

$$\mathbf{A}^T \mathbf{A}(\mathbf{A}^T \mathbf{y}) = \lambda(\mathbf{A}^T \mathbf{y}),$$

which means that λ is also an eigenvalue of $\mathbf{A}^T\mathbf{A}$. Finally, let us mention that in practice we do not need to find $\mathbf{x}$ and $\mathbf{y}$ independently, using (3.8) and (3.9). Once $\mathbf{y}$ is computed, then $\mathbf{x}$ can be obtained from (3.7) after fixing $\beta = 1$ as these scores need to be normalized anyway.

Combining all observations together, we are ready to define our next centrality measure. They are normalized so that the maximum centrality is equal to one.

Let $D = (V, E)$ be any directed graph on n nodes. Let λ be the unique leading eigenvalue of $\mathbf{A}\mathbf{A}^T$, where $\mathbf{A}$ is the adjacency matrix of graph G, and let $\mathbf{y}$ be the associated non-negative eigenvector; in particular, $\lambda\mathbf{y} = \mathbf{A}\mathbf{A}^T\mathbf{y}$. Let $\mathbf{x} = \mathbf{A}^T\mathbf{y}$. The **authority centrality** $c_a(v)$ and the **hub centrality** $c_h(v)$ of a node $v \in V$ are defined as follows:

$$c_a(v) = \frac{x(v)}{\max_{u \in V} x(u)} \quad \text{and} \quad c_h(v) = \frac{y(v)}{\max_{u \in V} y(u)}.$$

Computational Aspects

The eigenvector centrality measure requires finding a positive eigenvector that is associated with the leading eigenvalue of some matrix. Let us briefly discuss the most basic method for finding this eigenvector which is called **power iteration**. Suppose that our goal is to find the positive eigenvector associated with the leading eigenvalue of matrix $\mathbf{A}$ (from the discussion above we know that it exists). Let λ_1 be the leading eigenvalue of $\mathbf{A}$ and let λ_2 be the eigenvalue that has the second largest absolute value. Then, let us take any positive vector $\mathbf{c}^{(0)}$ and apply the following iteration:

$$\mathbf{c}^{(k+1)} = \frac{\mathbf{A}\mathbf{c}^{(k)}}{||\mathbf{A}\mathbf{c}^{(k)}||}, \quad \text{for any } k \in \mathbb{N} \cup \{0\}.$$

It is clear that $\mathbf{c}^{(k)}$ has norm equal to one and all positive entries. Moreover, it can be shown that the sequence converges geometrically in the rate of $|\lambda_2/\lambda_1|$. Let us also point out that in practice large graphs are usually sparse so the update can be done efficiently.

For Katz centrality, our goal is to find a solution $\mathbf{c}_\alpha$ to the following linear equation:

$$(\mathbf{I} - \alpha\mathbf{A}^T)\mathbf{c}_\alpha = \mathbf{1}.$$

Again, in practice, since large graphs are typically sparse, it is not efficient to find the inverse of $\mathbf{I} - \alpha\mathbf{A}^T$ so some matrix-free iterative solver is used instead. . The simplest approach is to iteratively calculate $\mathbf{c}^{(k+1)} = \alpha\mathbf{A}^T\mathbf{c}^{(k)} + \mathbf{1}$.

For PageRank we start with the formula

$$c_p = \frac{1-\alpha}{n}\left(I - \alpha\hat{A}^T\right)^{-1} 1$$

and rewrite it as follows:

$$\left(I - \alpha\hat{A}^T\right)c_p = \frac{1-\alpha}{n} 1.$$

Now, we see that $1 = Ec_p$, where E is a square matrix containing 1 everywhere, since entries of c_p are normalized so that they add up to one. It follows that

$$1 \cdot c_p = \left(\alpha\hat{A}^T + \frac{1-\alpha}{n} E\right)c_p.$$

Moreover, since the leading eigenvalue of $\alpha\hat{A}^T + \frac{1-\alpha}{n} E$ is equal to one, we can use the power iteration method to find c_p. We take $c_p^{(0)} = 1/n$ as a starting value and use the recurrence

$$c_p^{(k+1)} = \left(\alpha\hat{A}^T + \frac{1-\alpha}{n} E\right)c_p^{(k)}.$$

Note that this time we do not need to normalize $c_p^{(k)}$ as its norm is guaranteed to be 1. As usual, since $\hat{A}$ is assumed to be sparse for large n, the update in each iteration can be computed very fast.

Finally, for the HITS centrality measure, we initialize the algorithm with vectors $x^{(1)} = y^{(1)} = (1,1,\ldots,1)$. Then, during the $(k+1)$th step of the algorithm ($k \in \mathbb{N}$), we first perform the authority update rule:

$$x^{(k+1)} = A^T y^{(k)}/\|A^T y^{(k)}\|_\infty,$$

immediately followed by the hub update rule:

$$y^{(k+1)} = Ax^{(k+1)}/\|A^T x^{(k+1)}\|_\infty.$$

The values obtained from this process will eventually converge.

3.3   Distance Based Measures

Closeness Centrality

The first centrality measure we introduce in this section assumes that a node is central if it is close, on average, to the remaining nodes, as it can quickly interact with a typical or a random node. Potential applications include finding influential users on LinkedIn or identifying good intersections to build a hospital near by, but there are many similar scenarios where quick access to other nodes is crucial.

Since smaller average distance is more desirable, it makes sense to define the score of a node v as the reciprocal of the average distance from v to the remaining nodes, namely, $\sum_{u \in V} \text{dist}(v, u)/(n-1)$. For simplicity, we will assume in the definition below that graphs are unweighted but it is straightforward to generalize it to weighted graphs. We will come back to this afterwards. Recall that for undirected graphs, $\text{dist}(v, u)$ is simply the number of edges in a shortest path between v and u. For directed graphs, depending on the application at hand, it might be more natural to measure the average distance from node v (if nodes are central if they have an easy access to the rest of the graph) or the average distance to v (if the fact that the nodes have easy access to v make v central). We use the former variant below but it is trivial to adjust it to the latter.

Let $D = (V, E)$ be any strongly connected and unweighted directed graph on n nodes. Alternatively, let $G = (V, E)$ be any connected and unweighted graph on n nodes. The **closeness centrality** of a node $v \in V$ is defined as follows:

$$c_c(v) = \frac{n-1}{\sum_{u \in V} \text{dist}(v, u)}.$$

Let us first note that this measure is properly normalized for unweighted graphs. Indeed, trivially, $\text{dist}(v, v) = 0$ and $\text{dist}(v, u) \geq 1$ for $u \neq v$. So, in the extreme case, $\sum_{u \in V} \text{dist}(v, u) = n - 1$ and it holds for all nodes in the complete graph K_n or, for example, the center of the star $K_{1,n}$. For weighted graphs, $\text{dist}(v, u)$ is defined to be the minimum sum of weights taken over all paths from v to u (see (1.4)). As a result, we may stay with the same definition of closeness centrality for weighted graphs. However, the normalizing factor has to be adjusted in order to make sure that the centrality measure is always between zero and one.

Let us also mention that we need to assume that a graph is connected (or strongly connected in the case of directed graphs); otherwise, $\text{dist}(v, u)$ is not well defined if v and u belong to two different connected components. In order to generalize the closeness centrality to disconnected graphs, one may assume that $\text{dist}(v, u) = n$ whenever the two nodes belong to different components, the value just slightly larger than $n - 1$, the maximum distance two nodes can be apart from each other. This is true for both unweighted and weighted graphs; recall that it is assumed that the edge weights are always in $(0, 1]$.

Betweenness Centrality

The next centrality measure tries to capture the idea that a node that lies between many pairs of nodes has certain strategic and topological advantages and, as a result, is more influential compared to other nodes. Indeed, consider a network of roads in a large city such as Toronto. Our goal might be to model the behaviour of commuters; we know where they live and where they work. In order to identify the busiest intersection within the city, one might want to check which intersection lies on the largest number of shortest paths between commuters and their work places. Another natural application includes identifying the most crucial routers in a computer network but there are many other similar scenarios.

Though we will define this centrality measure for connected graphs, it is easy to generalize it to disconnected graphs.

Let $D = (V, E)$ be any strongly connected and unweighted directed graph on n nodes. Alternatively, let $G = (V, E)$ be any connected and unweighted graph on n nodes. For three distinct nodes $i, j, v \in V$, let $\ell(i, j)$ be the number of shortest paths from i to j, and let $\ell(i, j, v)$ be the number of shortest paths from i to j that include v. The **betweenness centrality** of a node $v \in V$ is defined as follows:

$$c_b(v) = \frac{1}{(n-1)(n-2)} \sum_{i \in V \setminus \{v\}} \sum_{j \in V \setminus \{v, i\}} \frac{\ell(i, j, v)}{\ell(i, j)}.$$

Note that the measure is properly normalized as there are $(n-1)(n-2)$ terms involved and each of them, $\ell(i, j, v)/\ell(i, j)$, is at most one. In order to generalize it to disconnected graphs, we may simply assume that $\ell(i, j, v)/\ell(i, j) = 0$ if there is no path from i to j (that is, $\ell(i, j) = 0$).

Note also that for weighted graphs, that often have highly heterogeneous edge weights, one should expect that $\ell(i, j) = 1$ for most of the pairs (i, j). Therefore, in practice, the above definition is slightly modified. One typical approach is to transform a graph into an unweighted counterpart in which an edge between two nodes is introduced if the weight of the edge linking them is greater than some universal threshold value. Another approach is to consider k-shortest paths between each pair of nodes (i, j). In the latter case, **Yen's algorithm** is a popular method to find the set of k paths of interest.

Efficiency and Delta Centrality

Let us start from the definition of graph efficiency. For simplicity, we define it for unweighted graphs but generalization to weighted graphs will be straightforward.

Let $D = (V, E)$ be any unweighted directed graph on n nodes. Alternatively, let $G = (V, E)$ be any unweighted graph on n nodes. For $i, j \in V$, the **efficiency** in the communication between node i and node j is defined as follows:

$$\delta(i, j) = \begin{cases} 1/\mathrm{dist}(i, j) & \text{if there is a path from } i \text{ to } j \\ 0 & \text{otherwise.} \end{cases}$$

The **graph efficiency** is defined as follows:

$$F = F(G) = \frac{1}{n(n-1)} \sum_{i \in V} \sum_{j \in V \setminus \{i\}} \delta(i, j).$$

Note that $\delta(i, j) \in [0, 1]$ for any two distinct nodes i and j. As a result, since the graph efficiency is simply the average value of $\delta(i, j)$ over all ordered pairs of nodes i, j with $i \neq j$, it is normalized so that $F(G) \in [0, 1]$. For weighted graphs, we simply replace the definition of $\mathrm{dist}(i, j)$ with (1.4) and properly normalize the graph efficiency.

Clearly, this graph parameter can be seen as a measure of the performance of the graph; the larger the value of $F(G)$, the more interconnected the nodes of graph G are. More importantly, once we have such a measure, we could use it to benchmark which nodes in the graph are central or important. Indeed, if after disconnecting node v from the graph, the efficiency goes down significantly compared to the same operation on other nodes, then it is a strong indication that v plays an important role within graph G.

Let us formalize these observations. For a given graph $G = (V, E)$ and a node $v \in V$, let $G_v = (V, E_v)$ be a subgraph of G with the edge set $E_v \subseteq E$ obtained from E by removing the edges incident with v. Informally, we may say that G_v is a graph G with node v **deactivated**. According to the centrality measure we are about to define, the most central node is the node which minimizes $F(G_v)$, the efficiency of G_v.

Let $D = (V, E)$ be any unweighted directed graph on n nodes. Alternatively, let $G = (V, E)$ be any unweighted graph on n nodes. The **Delta centrality (with respect to the graph efficiency)** of a node $v \in V$ is defined as follows:

$$c_\Delta(v) = \frac{F(G) - F(G_v)}{F(G)}.$$

Let us note that there are many ways to define the performance of a network and choosing which one to use should depend on the application at hand. In a very general setting, we say that $P(G)$ is a **quantity measuring the performance** of a graph $G = (V, E)$ if the following two properties hold:

a) $P(G) \in \mathbb{R}^+$ (provided that $|E| \geq 1$), and
b) for any $v \in V$ we have that $P(G_v) \leq P(G)$.

(A similar definition can be introduced for directed graphs.) Once the appropriate quantity measuring the performance is introduced, one can adjust the Delta centrality measure by simply replacing $F(G)$ with $P(G)$.

It is important to highlight that there are two possible scenarios that can take place. In both of them, we use exactly the same formula but the justification for it is quite different. Note that the goal could be to maximize $P(G)$ (when it measures some positive effect) in which case the nodes that correspond to large values of c_Δ are the ones that we should try to avoid deactivating as it results in drastic decrease of the graph performance. Alternatively, the goal could be to minimize $P(G)$ (when it measures some negative effect). This time the nodes of large values of c_Δ are the ones that we should try to deactivate. We will show one example of such customized quantity in Section 3.7.

It is straightforward to see that the graph efficiency $F(G)$ defined above is a quantity measuring the performance of G. In particular, it implies that the definition of the corresponding centrality measure above is a specification of this more general framework. Indeed, we already mentioned that it satisfies the first required property, namely, that $F(G) \in \mathbb{R}^+$ (unless graph G has no edges). To see that the second property is satisfied, note that after removing an edge from a graph, the distance between any pair of nodes can only increase and so the efficiency cannot increase.

3.4   Analyzing Centrality Measures

As already mentioned in the introduction, having centrality measures assigned to all nodes gives us flexibility and could be very useful. However, often our task is to identify a small set of central nodes of a given size, regardless of how far the next candidate that did not make into the set is. Hence, we often need to rank all the nodes based on a given centrality measure $c(v)$.

If all scores are unique, then there is a unique ranking. The most central node will be ranked one and the least central node will have rank n. If some nodes have exactly the same ranking, then we may use an **ordinal ranking** approach, that is, break ties using some additional rule. Alternatively, one may simply assign the same ranking to these nodes which can be done in several ways. The most popular among those are **standard competition ranking**, **modified competition ranking**, **dense ranking**, and **fractional ranking**.

Let us fix any graph $G = (V, E)$ or any directed graph $D = (V, E)$. Given a centrality measure $c\colon V \to [0, 1]$ that assigns a score $c(v)$ to all nodes v in a graph, the **ordinal ranking** of nodes is a permutation $r_o\colon V \to [n]$ such that $c(v) \geq c(w)$ whenever $r_o(v) < r_o(w)$. If $c(v) \neq c(w)$ for $v \neq w$, then the ranking is unique; otherwise, one may choose the ranking arbitrarily from the functions that satisfy a given criteria.

Out of the rankings that assign the same ranking to all the nodes that have the same score, one of the most common practice is to use the **fractional ranking**. Suppose that some nodes have exactly the same centrality measure. Then, all of them receive the same ranking that is equal to their average position. The benefit of this approach is that it preserves the average ranking of the corresponding ordinal ranking.

Let us fix any graph $G = (V, E)$ or any directed graph $D = (V, E)$. Given a centrality measure $c\colon V \to [0, 1]$ that assigns a score $c(v)$ to all nodes v in a graph, the **fractional ranking** of nodes is a function $r_f\colon V \to [1, n]$ such that $r_f(u) = \sum_{v \in S(u)} r_o(u)/|S(u)|$, where $S(u) = \{v \in V : c(v) = c(u)\}$.

Another common situation when rankings become useful is when we calculate several different centrality measures. In such cases, it is often important to identify nodes that have a high ranking with respect to several measures. Similarly, knowing that some node is ranked high with respect to one centrality measure while having low ranking under some other centrality measure might lead us to interesting and important insights regarding its role and position within the network.

Analysis of several centrality measures, at the same time, naturally lead to a question of how similar two centrality measures are for a given graph. Three most popular measures of correlation used in practice are **Pearson's** ρ, **Spearman's** r_s, and **Kendall's** τ. All of these measures are appropriately normalized such that they give a value in the $[-1, 1]$ interval, where 1 indicates a perfect association between the two centrality measures, 0 indicates no association, and -1 indicates that they are perfectly inverted.

The difference in interpretation of the three measures is the following. Pearson's ρ gives us information about *linear* relationship between data. Spearman's r_s measures the extent of presence of *monotone* relationships between data (or, equivalently, the presence of a linear relationship between ranks computed using original data). Finally, Kendall's τ can be roughly interpreted as follows: $(1 + \tau)/2$ is the probability that two observations taken randomly from both data sets have the same ordering of ranking. We will come back to this and provide more details once all coefficients are defined. Most statistical packages provide implementations of all three correlation measures.

Consider any two centrality measures, $c_1\colon V \to [0,1]$ and $c_2\colon V \to [0,1]$. The **Pearson's correlation coefficient** ρ of c_1 and c_2 is defined as follows:

$$\rho = \frac{\sum_{v \in V}(c_1(v) - \bar{c}_1)(c_2(v) - \bar{c}_2)}{\sqrt{\sum_{v \in V}(c_1(v) - \bar{c}_1)^2 \sum_{v \in V}(c_2(v) - \bar{c}_2)^2}},$$

where $\bar{c}_i = \frac{1}{n}\sum_{v \in V} c_i(v)$, $i \in \{1,2\}$, are the averages over all nodes.

Consider any two rankings, $r_1\colon V \to [1,n]$ and $r_2\colon V \to [1,n]$. The **Spearman's rank correlation coefficient** r_s of r_1 and r_2 is defined as follows:

$$r_s = \frac{\sum_{v \in V}(r_1(v) - \bar{r}_1)(r_2(v) - \bar{r}_2)}{\sqrt{\sum_{v \in V}(r_1(v) - \bar{r}_1)^2 \sum_{v \in V}(r_2(v) - \bar{r}_2)^2}},$$

where $\bar{r}_i = \frac{1}{n}\sum_{v \in V} r_i(v)$, $i \in \{1,2\}$, are the averages over all nodes.

Before we define Kendall's τ coefficient, we need to introduce a few definitions. Consider any two rankings, $r_1 : V \to [1,n]$ and $r_2 : V \to [1,n]$. For any pair u,v of nodes, we say that this pair is **concordant** if the ranks for both nodes agree, that is, if both $r_1(u) > r_1(v)$ and $r_2(u) > r_2(v)$; or if both $r_1(u) < r_1(v)$ and $r_2(u) < r_2(v)$. On the other hand, the pair is said to be **discordant** if $r_1(u) > r_1(v)$ and $r_2(u) < r_2(v)$; or if $r_1(u) < r_1(v)$ and $r_2(u) > r_2(v)$. If $r_1(u) = r_1(v)$ or $r_2(u) = r_2(v)$, then the pair is neither concordant nor discordant.

Consider any two rankings, $r_1\colon V \to [1,n]$ and $r_2\colon V \to [1,n]$. The **Kendall's rank correlation coefficient** τ of r_1 and r_2 is defined as follows:

$$\tau = \frac{n_c - n_d}{\sqrt{\left(\binom{n}{2} - n_1\right)\left(\binom{n}{2} - n_2\right)}},$$

where n_c is the number of concordant pairs and n_d is the number of discordant pairs. The terms n_1 and n_2 account for the presence of rank degeneracies in ranking r_1 and, respectively, r_2. If ranking r_j ($j \in \{1,2\}$) partitions V into ℓ_j groups, group $i \in [\ell_j]$ consists of nodes of the same rank and has k_i nodes, then $n_j = \sum_{i=1}^{\ell_j} \binom{k_i}{2}$.

Finally, as promised, let us make a connection to probability theory and the **Pearson's correlation coefficient** introduced in (1.2). Let X be a random variable uniformly distributed over the set $[n]$, that is, $\mathbb{P}(X = i) = 1/n$ for any $i \in [n]$. Then, simply

$$
\begin{aligned}
\rho &= \rho_{c_1(X),c_2(X)} \\
r_s &= \rho_{r_1(X),r_2(X)}.
\end{aligned}
$$

To see the connection to Kendall's rank correlation coefficient, let (X, Y) be a pair of random variables describing a uniform sample from the set $\{(x, y) : x, y \in [n], x \neq y\}$. Then,

$$
\tau = \rho_{\text{sgn}(c_1(X)-c_1(Y)),\text{sgn}(c_2(X)-c_2(Y))},
$$

where $\text{sgn}(x)$ is the **signum function** of a real number x defined as follows:

$$
\text{sgn}(x) = \begin{cases}
-1 & \text{if } x < 0, \\
0 & \text{if } x = 0, \\
1 & \text{if } x > 0.
\end{cases}
$$

So, indeed, all correlation coefficients can be reduced to the **Pearson's correlation coefficient** over different random variables.

3.5   Pruning Unimportant Nodes, k-cores

Before we define k-cores, we need to recall a definition of a subgraph. Let $G = (V, E)$ and $G' = (V', E')$ be any two undirected and unweighted graphs. A graph G' is a **subgraph** of G if $V' \subseteq V$ and $E' \subseteq E$. We say that a graph property $\mathcal{P}$ is **monotone** if for any subgraph G' of G we have the following: if G' has property $\mathcal{P}$, then G also has property $\mathcal{P}$. In other words, if a graph G' satisfies $\mathcal{P}$, then every graph G on the same set of nodes, which contains G' as a subgraph satisfies $\mathcal{P}$ as well. There are many natural monotone properties such as: a) the graph has minimum degree at least k, b) the graph is connected, c) the graph has diameter at most k, etc. Moreover, we say that a graph G' is a **maximal** subgraph of G satisfying some monotone property $\mathcal{P}$ if there is no graph $G'' = (V'', E'')$ with the same property $\mathcal{P}$ such that G' is a subgraph of G'', G'' is a subgraph of G, and $G'' \neq G'$. Informally speaking, we may say that there is no graph G'' "sandwiched" between G' and G which has property $\mathcal{P}$.

Extensions of k-cores to weighted, non-simple, or directed graphs are also developed but let us first concentrate on unweighted, undirected, and simple graphs.

Fix $k \in \mathbb{N}$. Let $G = (V, E)$ be any unweighted graph on n nodes. A k-**core** of a graph G is a maximal subgraph of G in which all nodes have degree at least k. It is easy to see that the k-core of G can be obtained by starting with G and repeatedly deleting all nodes of degree less than k.

Let us first note that k-cores are well defined, that is, there exists a unique graph that satisfies this definition. Indeed, if both $G_1 = (V_1, E_1)$ and $G_2 = (V_2, E_2)$ are graphs with minimum degree at least k, then the same is true for $G_{1 \cup 2} = (V_1 \cup V_2, E_1 \cup E_2)$. Hence, there is a unique maximal subgraph with this property.

Let us also note that the k-core of some graph G could be empty. For example, for any graph G, $(\Delta + 1)$-core is clearly empty, where $\Delta = \Delta(G)$ is the maximum degree of G. Moreover, all trees have empty 2-core, regardless of how large their maximum degree is. More importantly, for any $k_1 > k_2$ and any graph G, the k_1-core of G is a subgraph of the k_2-core of G. This justifies the following definition.

Let $G = (V, E)$ be any unweighted graph on n nodes. The **coreness** of a node $v \in V$ is equal to k if v belongs to the k-core but not to the $(k + 1)$-core.

The k-core decomposition identifies progressively internal cores and decomposes the graph layer by layer, revealing the structure of different "shells" from the outmost one to the most internal one. As a result, the coreness can then be used as a simple centrality measure. Note that each node has the coreness properly assigned as a natural number. This implies that the k-core decomposition yields a partition of V, the set of nodes.

However, the main problem here is that it is quite common that even the last non-empty core is still quite large. Indeed, for example, it is known that for any integer $k \geq 3$, a.a.s. the binomial random graph $\mathcal{G}(n, p)$ (see Section 2.3 for a definition and more) has either the empty core or it has linearly many nodes. Figure 3.1 presents the fraction of nodes that are part of the k-core of $\mathcal{G}(n, p)$ for $k \in [10]$, a property that holds a.a.s. Of course, these are asymptotic results but experiments confirm that similar conclusions can be derived even for small random graphs on, say, 1,000 nodes. Moreover, many real-world networks exhibit similar behaviour. As a result, there are typically many nodes with the highest score and so it is impossible to distinguish them and use the notion of coreness to identify the most central nodes. Having said that, k-cores may often be successfully used as a semi-supervised, preprocessing step to remove nodes that are not central. For example, "trimming" the graph to its 2-core is a very common first step in data analysis but larger values of k might also be considered. However, since it is not guaranteed that some important nodes are not removed during this procedure, it should be done with caution. Finally, let us state the obvious: this preprocessing step clearly provides a substantial improvement on the algorithmic complexity and so is, at least, worth considering.

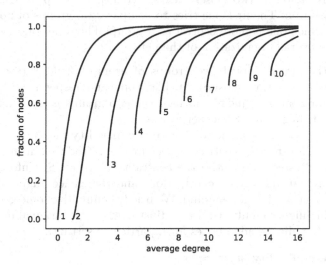

FIGURE 3.1
Asymptotic order of the k-cores of $\mathcal{G}(n,p)$ as a function of its (expected) average degree. All plots start at the point when the k-core becomes non-empty.

Let us briefly mention how one may generalize the notion of the k-core to weighted, non-simple, or directed graphs. The easiest thing to do is to simply ignore possible weights and directions of the edges and remove paralel edges and loops before finding the k-core. However, if a given application at hand requires paying attention to this additional information, then we suggest to use the following generalization instead.

Recall that for weighted graphs the degree of a node v is defined as $\deg(v) = \sum_{u \in V} a(v, u)$. This value can be used for the definition of the k-core, and so the generalization to these graphs is rather straightforward. The only difference to point out is that now the k-core may be defined for any $k \in \mathbb{R}_+$, which is not necessarily a natural number. For non-simple graphs, all parallel edges and loops incident to node v contribute to $\deg(v)$ (with loops typically contributing twice as much). Finally, for directed graphs, we may fix $k_{in}, k_{out} \in \mathbb{N}$ and define a k-core as a maximal subgraph in which all nodes have in-degree at least k_{in} and out-degree at least k_{out}. Alternatively, we may stay with only one parameter $k \in \mathbb{N}$ and insist on all nodes having total degree at least k.

3.6 Group Centrality and Graph Centralization

So far we were concerned with identifying central nodes, the nodes that play a key role within the graph. Let us finish the theoretical part of this chapter with a discussion of two closely related concepts: group centrality and graph centralization. The first one tries to compare two groups of nodes and then decides which one is more central as a group. The other one focuses on evaluating the overall organization of the network.

Suppose that we identified two groups of users of `Reddit` (social news aggregation and discussion website) who like Apple and, respectively, Samsung cellphones. Our goal is to judge which of the two groups is more influential and, as a result, might affect another group.

Our task is to assign some kind of **group centrality** measure to a group of nodes $S \subseteq V$. Most of the centrality measures we introduced in this chapter can be easily adjusted to quantify the centrality of group S. Unfortunately, there are many natural ways it can be done and the choice might be highly affected by a given application in mind. We briefly summarize some generalizations below, highlighting only required adjustments to the original definitions (see the corresponding shaded boxes for a particular setting).

- The **degree centrality** of set S:

$$c_d(S) = \frac{|N(S)|}{n - |S|},$$

  where $N(S) = \{u \in V \setminus S : uv \in E \text{ for some } v \in S\}$.

- Each of the remaining **matrix based measures** (**Eigenvector, Katz, PageRank, HITS**) of set S:

$$c(S) = \sum_{v \in S} c(v).$$

- The **closeness centrality** of set S:

$$c_c(S) = \frac{n - |S|}{\sum_{u \in V} \text{dist}(S, u)},$$

  where $\text{dist}(S, u) = \min_{v \in S} \text{dist}(v, u)$ is the distance from u to the closest node in S.

- The **betweenness centrality** of set S:

$$c_b(S) = \frac{1}{(n - |S|)(n - |S| - 1)} \sum_{i \in V \setminus S} \sum_{j \in V \setminus (S \cup \{i\})} \frac{\ell(i, j, S)}{\ell(i, j)},$$

where $\ell(i,j)$ is the number of shortest paths from i to j (this part has not changed) and $\ell(i,j,S)$ is the number of shortest paths from i to j that include at least one node from S.

- In order to generalize the **Delta centrality** from individual nodes to groups of nodes, we simply replace G_v with G_S, where $G_S = (V, E_v)$ is a subgraph of G with the edge set $E_S \subseteq E$ obtained from E by removing all edges incident with some node in S.

Finally, let us point out that these generalizations can be easily used when we already have a number of groups identified and all we need to do is to rank them. But what if we do not have candidates in mind and our goal is to find a group of nodes that is maximally central? This is clearly an important task but much harder than the one we discussed above. There are $\binom{n}{k}$ groups of nodes of size k which is too large to investigate even for small values of k. As a result, some heuristic search algorithms have to be applied which do not guarantee to find the best group but aim for a reasonable outcome instead. These problems are out of scope of this book. Another important and related problem that is too challenging to be included here is concerned with the question of how to change the structure of a given graph in order to increase a centrality of a given group of nodes.

Let us now move to the second problem. Our goal now is to try to understand how centrality is distributed over the nodes of the graph. For example, we would like to distinguish graphs in which all nodes play similar role with no clear influencers from graphs which consist of some small group of nodes with large centrality. The graph centralization is a graph parameter that, for a given centrality measure $c_G : V \to [0,1]$, tries to investigate how that measure is distributed among nodes in the graph by associating a single number to a given graph $G = (V, E)$. Large values should imply that the measure is concentrated on a small set of central nodes; on the other hand, small values should indicate that the measure is almost uniformly distributed among the nodes of the graph.

One natural option for this graph parameter is to see how far a score of a typical node is from the score of the top ranked node, that is, to consider

$$c(G) = \frac{1}{n} \sum_{v \in V} \left(\max_{u \in V} c_G(u) - c_G(v) \right) = \max_{u \in V} c_G(u) - \frac{1}{n} \sum_{v \in V} c_G(v).$$

Note that the formula for $c(G)$ is simplified in some special cases. For example, if a centrality measure is standardized so that the sum of all centralities is equal to 1 (as in the case of PageRank), then we get that $c(G) = \max_{u \in V} c_G(u) - 1/n$; so we are effectively concentrating on the maximum value. Another popular standardization is to make the largest value of centrality equal to 1 (as in the case of eigenvector centrality), in which case we have $c(G) = 1 - \sum_{v \in V} c_G(v)/n$; this time, we are effectively looking at the negation of the average centrality.

If the centrality measure is normalized, that is, $c_G(v) \in [0,1]$, then we clearly also have that $c(G) \in [0,1]$. However, it is always the case that $c(G) <$ $\max_{u \in V} c_G(u)$ and so $c(G) < 1$, provided that the measure is normalized. Indeed, if the maximum is equal to one, then the average cannot be equal to zero. Hence, $c(G)$ is typically scaled so that it is equal to one for some extremal graph G' on the same set of nodes. As a result, we get the following definition.

For a given centrality measure, let $c_G \colon V \to [0,1]$ be the measure applied to a graph $G = (V,E)$. The **centralization** of graph G, with respect to the selected centrality measure, is defined as follows:

$$C(G) = \frac{\max_{u \in V} c_G(u) - \sum_{v \in V} c_G(v)/n}{\max_{G'=(V,E')} \left(\max_{u \in V} c_{G'}(u) - \sum_{v \in V} c_{G'}(v)/n \right)},$$

where the maximum in the denominator is taken over all graphs on the same set of nodes. The centralization of a directed graph $D = (V,E)$ is defined analogously.

Let us note that the denominator in the definition of $C(G)$ might be challenging to find. However, this is usually not a problem as in a typical scenario we have a few graphs to compare and so one may simply compare the ratios between the corresponding centralizations and the denominator is irrelevant, provided that the orders of the two graphs are comparable. Finally, despite the fact that this graph parameter could be useful as it provides a single number that can be assigned to a graph, many important properties are clearly lost. Hence, it is often recommended to consider a histogram of the distribution of the centrality measure we are interested in as it provides more detailed picture.

3.7   Experiments

The examples in this section are based on the USA Airport dataset at Kaggle that is available from their website[1]. From this dataset, we extracted a directed, weighted graph consisting of all flights in 2008 between US airports; the edge weights correspond to the total number of passengers transported between the two nodes. The graph is directed as we could distinguish flights from A to B and from B to A.

---

[1] www.kaggle.com/flashgordon/usa-airport-dataset

The graph consists of 464 nodes and 12,000 directed, weighted edges. For the nodes, we used the airport IATA codes as their labels. The data is imported from two files. First, we read a file that contains all the edges—see Table 3.2. The second file contains some properties of the nodes (the airports)—see Table 3.3.

TABLE 3.2
Weighted edges for the US Airport dataset.

orig_airport	dest_airport	total_passengers
SFO	LAX	1442105
LAX	SFO	1438639
MCO	ATL	1436625
ATL	MCO	1424069
LAX	JFK	1277731
...	...	...

TABLE 3.3
Node features for the US Airport dataset.

airport	lon	lat	state	city
ABE	−75.441	40.652	PA	Allentown
ABI	−99.682	32.411	TX	Abilene
ABQ	−106.609	35.040	NM	Albuquerque
ABR	−98.422	45.449	SD	Aberdeen
ABY	−84.195	31.535	GA	Albany
...	...	...	...	...

For the purpose of having a nicer visualization, we consider a subgraph obtained from the US Airport dataset consisting only of nodes and edges within the state of California. Details on how to build the dataset as well as the analysis done in this section can be found in the notebooks available on-line.

In Figure 3.4, we present the whole subgraph using the latitude and longitude of each node. From those plots, it is already visually clear that some nodes are more "central" than others, with a large number of edges coming in and out of them. In Figure 3.4(a), we highlight (in black) the three airports with the largest centrality values with respect to most of the measures we considered earlier. Not surprisingly, those central nodes represent LAX (Los Angeles), SFO (San Francisco), and SAN (San Diego). In Figure 3.4(b), we highlight the airports that have the largest (in black) and the smallest (in light grey) coreness numbers.

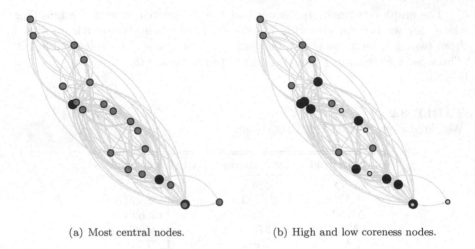

(a) Most central nodes.   (b) High and low coreness nodes.

FIGURE 3.4
The California subgraph using geographical layout. In (a), the top 3 most central nodes are highlighted. In (b), nodes with large (black) and small (light grey) coreness are presented.

In Table 3.5, we show the top ranked nodes, sorted with respect to their degree centrality. For all centrality measures, we used the edge weights counting the number of passengers between the corresponding airports. One clear observation is that the centrality measures seem to be highly correlated, showing a clear hierarchy. For betweenness, however, we see an enormous dominance of LAX and SFO, which indicates that for most connecting flights, one would need to go through one of those airports. Closeness centrality is similar for all airports, which is an indication of a very tightly connected graph in which it is possible to go between most airports in just one or a few hops (we investigate the centralization of this graph against other states later in this section). In Table 3.6, we show the airports with the lowest degree centrality. The corresponding numbers are all very low, except for closeness as we discussed, but two airports VIS (Visalia) and MCE (Merced) have much smaller values. What could explain that? We will come back to this shortly.

In this chapter, we saw several definitions of correlation between measured quantities. In Table 3.7, we show the Kendall's correlation coefficient for the measures presented earlier. While the correlation values are generally high, we see slightly lower values for betweenness and closeness in comparison to other pairs of centrality measures. For betweenness, we already saw that this is influenced by the dominance of two major airports while for closeness all values are similar.

We now consider the coreness (or the core number) for each airport. In this experiment, we treat the graph as undirected and unweighted, so there is a link

TABLE 3.5
Top ranked nodes sorted with respect to degree centrality.

airport	degree	pagerank	authority	hub	between	closeness
LAX	0.117	0.215	1.000	1.000	0.507	0.318
SFO	0.090	0.173	0.970	0.907	0.362	0.318
SAN	0.079	0.124	0.689	0.726	0.014	0.292
OAK	0.047	0.073	0.474	0.441	0.030	0.288
SJC	0.042	0.067	0.416	0.390	0.047	0.300

TABLE 3.6
Bottom ranked nodes sorted with respect to degree centrality.

airport	degree	pagerank	authority	hub	between	closeness
CEC	0.000	0.008	0.005	0.005	0.001	0.266
IPL	0.000	0.007	0.004	0.004	0.000	0.250
VIS	0.000	0.045	0.000	0.000	0.000	0.048
MCE	0.000	0.045	0.000	0.000	0.000	0.048
NZY	0.000	0.007	0.000	0.000	0.000	0.250

TABLE 3.7
Kendall's correlation coefficients between various centrality measures for the California airports graph.

	degree	pagerank	authority	hub	between	closeness
degree	1.000	0.774	0.989	0.974	0.708	0.812
pagerank	0.774	1.000	0.764	0.748	0.578	0.590
authority	0.989	0.764	1.000	0.980	0.697	0.801
hub	0.974	0.748	0.980	1.000	0.699	0.821
between	0.708	0.578	0.697	0.699	1.000	0.855
closeness	0.812	0.590	0.801	0.821	0.855	1.000

between two airports so long as there is at least one directed edge. The highest core value for this graph is 13, with 9 airports present in the last non-empty core, namely, SFO, LAX, SAN, OAK, SNA, SJC, SMF, FAT, and SBA. There are 5 airports with core number 2 or less, including MCE and VIS, and the remaining 8 airports have core number between 4 and 11. In Figure 3.8(a), we display the nodes that belong to the 13-core in black, and the low core airports in light grey, as we did in Figure 3.4(b). However this time, rather than using geographical layout, we use a force-directed layout (see Section 6.5 for more details on that). It is now clear that the nodes that belong to the 13-core are the most central. We also see that two airports are not part of the main connected component; not surprisingly, those are the MCE and VIS airports we already noticed. In Figure 3.8(b), we show the same plot but this time additionally with the (undirected unweighted) closeness centrality values. There is a clear pattern: larger values for the central nodes, slightly lower for the non-central nodes that are still part of the main connected component, and low values for the disconnected pair.

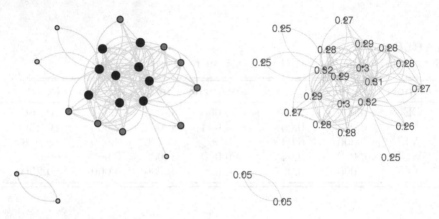

(a) High (black) and low (light grey) undi-   (b) Closeness centrality values (undirected
rected coreness nodes.   and unweighted).

FIGURE 3.8
The California airport graph shown with a force directed layout. The dense region and an isolated pair now appear more clearly.

In Figure 3.9, we plot the median centrality measures grouped by coreness values. We see a clear dominance of the high-core nodes for all measures except closeness centrality for which the values are more similar to each other.

In our next example we discuss delta centrality using a non-standard efficiency measure where we consider a simple pandemic spread model. Let us assume that the pandemic starts at exactly one airport selected uniformly at random from all the airports. Then, the following rules for spreading are applied: (i) in a given airport pandemic lasts only for one round and (ii) in the next round, with probability α, the pandemic spreads independently along the flight routes to the destination airports for all connections starting from this airport. Airports can interact with the pandemic many times, and the process either goes on forever or the pandemic eventually dies out. Our goal is to find the expected number of times a given airport interacted with the pandemic, which amounts to the sum over all airports of the expected number of times this airport has the pandemic. This is directly related to attenuation factor

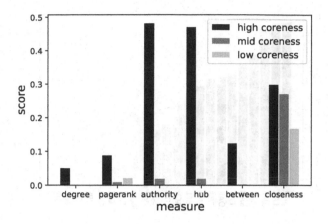

FIGURE 3.9
Median centrality measures for nodes grouped by their coreness: high (coreness=13), low (coreness ≤ 2) and mid-values.

we discussed above and can be calculated as follows:

$$P(G) = \mathbf{1}^T(\mathbf{I} - \alpha\mathbf{A}^T)^{-1}\mathbf{1}/n,$$

where $\mathbf{A}$ is the adjacency matrix, n is the number of nodes and we fix $\alpha = 0.1$. (A careful reader might realize that the presented formula is not completely correct. It is possible that at some stage of the pandemic process an airport gets infected two or more times in the same round, and we are double counting such cases. However, the probability of such a situation is small so we ignore it to keep formulas simple. An implementation of a simulation that correctly handles these situations is left as a programming exercise to the reader.) We use the above measure to compute delta centrality for each airport. Let us note that it has a simple but important interpretation: it simply computes by what percent the expected pandemic exposure is reduced when a selected airport is closed.

We computed that in the original graph $P(G) = 6.9403$, so we expect that a pandemic will be present in some airport almost 7 times when $\alpha = 0.1$. The results for delta centrality are shown in Figure 3.10, where we use the same shades of grey as with the coreness. We see that SFO and LAX stand out as having the highest positive impact when removed.

In order to illustrate group centrality and centralization, we now consider the entire US airport graph. For group centrality, we compare delta centrality (using standard efficiency) when removing all edges from each state in turn. In Table 3.11, we show the highest and the lowest scoring states while in neighbouring Table 3.12, we show the states with the largest and the smallest

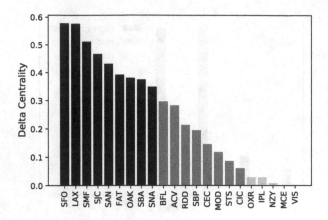

FIGURE 3.10
Delta centrality for a simple pandemic spread model.

un-normalized centralization using the PageRank centrality measure. We only considered states with more than 5 airports for the latter, in order to clearly illustrate the structure of graphs with high or low centralization. The state with largest centralization (MI) has one central hub airport, namely Detroit (DTW), while the state with smallest centralization (ND) does not have any. We plot those two subgraphs in Figure 3.13. We also note that California is not amongst the states with highest centralization, which agrees with our previous observation that closeness centrality is quite uniform due to high connectivity between (most) airports in that state.

TABLE 3.11
States with the largest and the smallest group delta centrality in the US Airport graph.

State	Delta centrality
TX	0.159
CA	0.134
FL	0.113
...	...
DE	0.008
RI	0.006
NH	0.005

TABLE 3.12
States with the largest and the smallest Pagerank Centralization in the US Airport graph.

State	Centralization
MI	0.373
GA	0.361
NC	0.358
...	...
NE	0.077
AR	0.067
ND	0.061

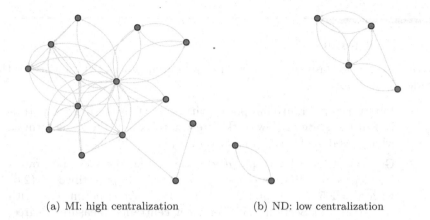

(a) MI: high centralization   (b) ND: low centralization

FIGURE 3.13
States with more than 5 airports with the largest (a) and the smallest (b) Pagerank centralization.

3.8   Practitioner's Corner

While all matrix-based centrality measures we discussed above are often highly correlated, PageRank is typically recommended (instead of, for example, eigenvector centrality or Katz centrality) for very large weighted and directed graphs, as it is the most robust method from a computational point of view. Clearly, in cases where the distinction between hubs and authorities is to be made, the HITS algorithm is a good choice. After all, it is designed for that purpose. If a tailor-made notion of an efficiency of a graph can be naturally defined, then delta-centrality provides a nice and flexible way to rank the nodes.

For weighted graphs, it is particularly important to pay attention to the type of edge weights. For matrix-based centrality measures, such weights should amount to some notion of similarity between the nodes. On the other hand, for betweenness they should map to some notion of a distance. In examples such as the airport graph that we have discussed above, it is often simpler to ignore the weights for betweenness measure and use hop counts as the distance between nodes instead.

Finally, pruning a graph by keeping the k-core is often useful in many applications as it allows the data scientist to concentrate on a smaller graph consisting of the most interesting nodes anyway. In particular, restricting to the 2-core will remove "dangling" trees that are typically uninteresting.

3.9 Problems

In this section we present a collection of potential practical problems for the reader to attempt.

1. Find the top 5 ranked airports (with respect to the degree centrality) in the state of New York. Present results in the same form as what was done for the state of California—Table 3.5.

2. Generate Chung-Lu graph $\mathcal{G}(\mathbf{w})$ on $n = 10{,}000$ nodes and power-law degree distribution using the set of weights prescribed by (2.6) with $\gamma = 2.5$, $\delta = 1$, and $\Delta = \sqrt{n} = 100$. (We experimented with this model in Section 2.5.) For each centrality measure (degree, PageRank, authority, hub, between, closeness) compute $f(k)$, the average centrality measure over all nodes of degree k. Plot $f(k)$ as a function of k. Is there any visible correlation between a given centrality measure and the degree of a node?

3. The original airport dataset (464 nodes, 12,000 directed weighted edges) is not strongly connected but it is almost weakly connected. There are two weak components: the giant one consisting of all but two airports, and the small one with two airports that only link to each other.Remove the two nodes to make the graph weakly connected. For each centrality measure (degree, PageRank, authority, hub, between, closeness) perform the following experiment: sort all nodes with respect to a given centrality measure. How many nodes with the largest score do we need to remove so that the graph is no longer weakly connected? Compare this with the number of nodes with the smallest score that need to be removed.

4. Generate a family of binomial random graphs $\mathcal{G}(n, p)$ with $n \in \{1250, 2500, 5000, 10000, 20000\}$ and with expected degree pn from the set $\{1, 2, 4, 8, 16\}$. In order to make these graphs connected, form a giant component by removing small components. For each centrality measure (degree, PageRank, authority, hub, between, closeness) estimate the time required to compute it. What can you say about the scalability of the algorithms under the implementation you have used? Try to estimate the complexity of the algorithms used.

5. From the airport graph, generate a state-to-state graph by collapsing every node in a state to a single node, adding the edge weights and removing loops (edges between airports in the same state). In `igraph`, the function `contract_vertices()` can be used to do this. Let G be this new graph with 51 nodes (the 50 states plus DC).

   (a) Is G weakly connected? Strongly connected?

(b) Which state has the most incoming passengers? The most departing passengers?

(c) Which pair of two states, x and y, have the most passengers travelling from x to y (directed weight)?

(d) Recreate Table 3.5 for graph G. Which states have the highest degree centrality? Betweenness centrality?

3.10 Recommended Supplementary Reading

There is a vast literature on centrality measures in networks. For a comprehensive overview we refer the reader to the following book and the survey:

- A.N. Langville, C.D. Meyer. *Googles PageRank and beyond: The science of search engine rankings*. Princeton University Press, 2011.

- P. Boldi, S. Vigna. "Axioms for centrality". *Internet Mathematics* **10** (2014), 222–262.

Here is a list of papers on specific measures:

- P. Bonacich, P. Lloyd. "Eigenvector-like measures of centrality for asymmetric relations". *Soc. Networks* **23** (2001), 191–201. (Eigenvector Centrality and Katz Centrality)

- S. Brin and L. Page. "The anatomy of a large-scale hypertextual Web search engine". *Comput. Networks ISDN Systems* **30(1-7)** (1998), 107–117. (PageRank Centrality)

- J.M. Kleinberg. "Authoritative sources in a hyperlinked environment". *J. ACM* **46** (1999), 604–632. (HITS)

- L. Freeman. "Centrality in social networks: conceptual clarification". *Soc. Networks* **1** (1979), 215–239. (Closeness Centrality)

- L. Freeman. "A set of measures of centrality based on betweenness". *Sociometry* (1977). (Betweenness Centrality)

- V. Latora, M. Marchiori. "A measure of centrality based on network efficiency". *New J. Phys.* **9** (2007), 188. (Delta Centrality)

- M.G. Everett, S.P. Borgatti. "The centrality of groups and classes". *J. Math. Sociol.* **23** (1999), 181–201. (Group Centrality)

- S. Wasserman, K. Faust. *Social Network Analysis: Methods and Applications*. Vol. 8. Cambridge University Press, 1994. (Graph Centralization)

4

Degree Correlations

4.1 Introduction

Consider a scientific collaboration network in which nodes are, say, mathematicians and edges are co-authorships. In other words, a mathematician v is adjacent to mathematician u in this graph if they published at least one research paper together. It is an undirected, scale-free network with the degree distribution following a power law with an exponential cutoff. Most mathematicians are sparsely connected while a few of them are highly connected. This network has an interesting property that we aim to investigate in this chapter, namely, hubs tend to be adjacent to other hubs and low-degree nodes tend to interact with another low-degree nodes. It is perhaps not too surprising. After all, it is more likely that two Fields medalists or Abel Prize winners would find a project of common interest rather than them collaborating with a first year Ph.D. student entering the world of mathematics. This is not to say that it is impossible. Indeed, the mathematical community seems to be quite open with the famous example of Paul Erdős who used to work with everyone who had "her or his brain open." However, on average, there are less interactions between members of this community of such type.

On the other hand, the web graph, a network in which nodes correspond to web pages and directed edges correspond to hyperlinks between them, exhibits the opposite property. In this network, high-degree nodes tend to be adjacent to low-degree nodes. Similar behaviour can be found in the protein interaction network of yeast. This time, each node corresponds to a protein and two proteins are linked if there is experimental evidence that they can bind to each other in the cell.

This chapter is structured as follows. We first give an intuition behind two possible types of correlations one might expect in networks: assortativity and disassortativity (Section 4.2). Then, we provide formal tools to distinguish the two (Section 4.3). The following two sections discuss some natural ways for correlations to appear (Section 4.4) and ways to generalize these concepts to directed graphs (Section 4.5). Next, we discuss some important implications (Section 4.6). As usual, we finish the chapter with experiments (Section 4.7) and provide some tips for practitioners (Section 4.8).

DOI: 10.1201/9781003218869-4

4.2 Assortativity and Disassortativity

For simplicity, let us assume that $G = (V, E)$ is any unweighted graph on n nodes. We will explain how to deal with directed graphs later on, in Section 4.5. However, it seems that there is no natural generalization of these concepts to weighted graphs. Recall that the degree distribution d_ℓ of a graph is defined to be the fraction of nodes with degree ℓ, that is, $d_\ell = n_\ell/n$, where n_ℓ is the number of nodes of degree ℓ (see Section 1.8 for more details).

The goal of this section is to introduce the notation and basic definitions that will be used to describe the properties of correlated graphs. The most general way to do that is to randomly select one edge and then investigate the properties of the two associated endpoints. That leads us to the following definitions.

Let $G = (V, E)$ be any unweighted graph on n nodes and $m \geq 1$ edges, and let $\ell_1, \ell_2 \in \mathbb{N}$. Let $e \in E$ be an edge selected uniformly at random from E. Then, a fair coin is tossed to identify the first end of e, node u; the other end is node v.

- q_{ℓ_1} is the probability that u has degree ℓ_1.

- $p(\ell_1, \ell_2)$ is the **joint probability** that u has degree ℓ_1 and v has degree ℓ_2.

- Assuming that $q_{\ell_1} \neq 0$ (which is equivalent to $d_{\ell_1} \neq 0$), $p(\ell_2|\ell_1)$ is the **conditional probability** that v has degree ℓ_2, given that u has degree ℓ_1.

Let us first note that in some references q_{ℓ_1} and, consequently all other variables, are defined in terms of the *remaining degree* of a given node, that is, its degree *minus one* (the degree after removing the edge that was used to get to node of degree ℓ_1).

Let us also note that the conditional probability $p(\ell_2|\ell_1)$ is well defined only if nodes of degree ℓ_1 are present in the graph, that is, when $d_{\ell_1} \neq 0$. Hence, every time we use $p(\ell_2|\ell_1)$ and we do not explicitly mention this assumption, it is implicitly assumed anyway. However, for convenience let us define $\mathcal{D} = \{\ell \in \mathbb{N} : d_\ell > 0\}$, that is, $\mathcal{D}$ is the set of all positive node degrees in G. Clearly, $\max \mathcal{D} = \Delta$, where Δ is the maximum degree of a graph G. Note also that

$$q_{\ell_1} = \frac{\ell_1 \cdot n_{\ell_1}}{\sum_{\ell \in \mathcal{D}} \ell \cdot n_\ell} = \frac{\ell_1 \cdot n_{\ell_1}}{2m} = \frac{\ell_1 \cdot n_{\ell_1}/n}{2m/n} = \frac{\ell_1 \cdot d_{\ell_1}}{\langle k \rangle},$$

where $\langle k \rangle = 2m/n$ is the average degree in G. Indeed, there are $\ell_1 \cdot n_{\ell_1}$ "half-edges" associated with nodes of degree ℓ_1 and $2m$ "half-edges" in total. Moreover, both joint and conditional probabilities completely characterize the way degrees are correlated. Indeed, we will show that the two probabilities contain the same information but there are some important differences. For example, note that $p(\ell_1, \ell_2)$ is symmetric whereas $p(\ell_2 | \ell_1)$ does not have to be. Using the following formula from probability theory

$$P(A|B) = \frac{P(A \cap B)}{P(B)},$$

we immediately get that for $\ell_1 \in \mathcal{D}$

$$p(\ell_2 | \ell_1) = \frac{p(\ell_1, \ell_2)}{q_{\ell_1}}$$

or, alternatively, that

$$p(\ell_1, \ell_2) = q_{\ell_1} \cdot p(\ell_2 | \ell_1). \tag{4.1}$$

Because of this relationship, we may restrict ourselves to one of them. We will mostly concentrate on the **degree correlation matrix**

$$P = \big(p(\ell_1, \ell_2)\big)_{\ell_1, \ell_2 \in [\Delta]}$$

consisting of joint probabilities. Note that matrix P is an $\Delta \times \Delta$ matrix, where Δ is the maximum degree. In simple graphs, trivially $\Delta \le n-1$ but very often it is much smaller than that.

Let us highlight some basic properties of this matrix. Observe first that $p(\ell_1, \ell_2) = p(\ell_2, \ell_1)$ and $\sum_{\ell_1 \in \mathcal{D}, \ell_2 \in \mathcal{D}} p(\ell_1, \ell_2) = 1$. If $\ell_1 \ne \ell_2$, then $2m \cdot p(\ell_1, \ell_2)$ is the number of edges (recall that the graph is undirected) between nodes of degree ℓ_1 and nodes of degree ℓ_2. On the other hand, the interpretation of the entries on the diagonal of matrix P is slightly different since the number of edges between nodes that both have degree ℓ is $m \cdot p(\ell, \ell)$.

Using (4.1) we get that for $\ell_1 \in \mathcal{D}$

$$\sum_{\ell_2 \in [\Delta]} p(\ell_1, \ell_2) = q_{\ell_1} \cdot \sum_{\ell_2 \in [\Delta]} p(\ell_2 | \ell_1) = q_{\ell_1}.$$

If $\ell_1 \notin \mathcal{D}$, then clearly $\sum_{\ell_2 \in [\Delta]} p(\ell_1, \ell_2) = q_{\ell_1} = 0$. Hence, q_{ℓ_1} is the marginal probability associated with the joint probability $p(\ell_1, \ell_2)$. The last property we want to mention is known as the **detailed balance condition**:

$$\ell_1 \cdot p(\ell_2 | \ell_1) \cdot d_{\ell_1} = \ell_1 \frac{p(\ell_1, \ell_2)}{q_{\ell_1}} d_{\ell_1} = \ell_1 \frac{p(\ell_1, \ell_2)}{\ell_1 \cdot d_{\ell_1} / \langle k \rangle} d_{\ell_1} = p(\ell_1, \ell_2) \cdot \langle k \rangle,$$

which implies that

$$\ell_1 \cdot p(\ell_2 | \ell_1) \cdot d_{\ell_1} = \ell_2 \cdot p(\ell_1 | \ell_2) \cdot d_{\ell_2}. \tag{4.2}$$

This equality has a natural interpretation. After multiplying both sides by n, the left hand side counts the number of edges from nodes of degree ℓ_1 to nodes of degree ℓ_2 (we may think of these edges as being oriented). The right hand side does the same but counts these edges the other way (the edges change their orientations).

Now, we are ready to give a formal definition of correlated and uncorrelated networks.

Let $G = (V, E)$ be any unweighted graph on n nodes and $m \geq 1$ edges. We say that G is **uncorrelated** (or has no degree-degree correlation) if the conditional probability $p(\ell_2|\ell_1)$ does not depend on $\ell_1 \in \mathcal{D}$. Otherwise, we say that G is **correlated**.

In order to find an explicit formula for the conditional probability in the case of uncorrelated graphs, let us fix $\ell_i \in \mathcal{D}$ and use (4.1) to observe that

$$1 = \sum_{\ell_j \in \mathcal{D}} p(\ell_j|\ell_i) = \sum_{\ell_j \in \mathcal{D}} p(\ell_i|\ell_j) \cdot q_{\ell_j}/q_{\ell_i},$$

so as $p(\ell_i|\ell_j)$ is assumed to be constant (independent of j) in an uncorrelated graph we get:

$$1 = p(\ell_i|\ell_j) \sum_{\ell_j \in \mathcal{D}} q_{\ell_j}/q_{\ell_i}.$$

A final note is that $\sum_{\ell_j \in \mathcal{D}} q_{\ell_j} = 1$ and thus $p(\ell_i|\ell_j) = q_{\ell_i}$. Combining this with (4.1) we get the following simple but important observation. It enables us to benchmark graphs and recognize positive and negative correlations. For that reason we will use $\hat{p}$ as a benchmark for uncorrelated probabilities whereas we keep p for a specific graph to test.

If G is an **uncorrelated** graph, then the **conditional** and **joint probabilities** reduce to the following expressions:

$$\hat{p}(\ell_2|\ell_1) = q_{\ell_2} \qquad \text{and} \qquad \hat{p}(\ell_1, \ell_2) = q_{\ell_1} \cdot q_{\ell_2}.$$

Now, we are ready to define three types of degree-degree correlations. The definition below is not precise but hopefully provides a good intuition behind these concepts. In the next section, we will provide much better tools to distinguish these types of degree-degree correlations.

A network is **neutral** if it looks like one would wire edges at random, regardless of what degrees the corresponding end nodes have. For such graphs, $p(\ell_1, \ell_2) \approx \hat{p}(\ell_1, \ell_2) = q_{\ell_1} \cdot q_{\ell_2}$ for most values of $\ell_1, \ell_2 \in \mathcal{D}$. In **assortative** graphs, hubs tend to be adjacent to each other and avoid linking to small degree nodes. On the other hand, in such graphs, small degree nodes tend to be connected to other small degree nodes. Finally, if hubs tend to be adjacent to small degree nodes, then such **disassortative** networks will exhibit a "hub-and-spoke" character.

4.3 Measures of Degree Correlations

As pointed out in the previous section, the degree correlation matrix contains complete information about the degree correlations present in a given graph. Unfortunately, it is difficult to interpret it as the entries $p(\ell_1, \ell_2)$ in the matrix are typically small and for empirical graphs often do not change monotonically as one changes ℓ_1 or ℓ_2. Still, they typically exhibit some clear trend in the way they change. As a result, it is much easier to understand the correlations by looking at the average degree of the neighbours of all nodes of degree ℓ, a function that is much more stable and thus easier to interpret. This leads us to the following definition.

Let $G = (V, E)$ be any unweighted graph on n nodes and $m \geq 1$ edges, and let $\ell \in \mathcal{D}$ (that is, $d_\ell \neq 0$). The **degree correlation function** is defined as follows:

$$k_{nn}(\ell) = \sum_{\ell' \in [\Delta]} \ell'\, p(\ell'|\ell),$$

where $p(\ell'|\ell)$ is the conditional probability that a node of degree ℓ has a neighbour of degree ℓ'.

Before we try to use the degree correlation function to distinguish assortative networks from disassortative ones, let us understand its behaviour for neutral networks. This understanding will be useful for benchmarking correlated graphs. Since for uncorrelated graphs $p(\ell'|\ell) = \hat{p}(\ell'|\ell) = q_{\ell'}$, we get that for such graphs, the degree correlation function is equal to

$$\hat{k}_{nn}(\ell) = \sum_{\ell' \in \mathcal{D}} \ell'\, \hat{p}(\ell'|\ell) = \sum_{\ell' \in \mathcal{D}} \ell' q_{\ell'} = \sum_{\ell' \in \mathcal{D}} \ell' \frac{\ell' \cdot d_{\ell'}}{\langle k \rangle} = \frac{\langle k^2 \rangle}{\langle k \rangle}, \qquad (4.3)$$

where $\langle k \rangle$ and $\langle k^2 \rangle$ are the average degree and, respectively, the second moment. As a result, the degree correlation function does not depend on ℓ but only on the global properties of the graph. Hence, for neutral networks one expects $k_{nn}(\ell)$ to be close to a constant function, namely, to the constant $\langle k^2 \rangle / \langle k \rangle$.

For instance, let us consider the binomial random graph $\mathcal{G}(n, p)$. Clearly, it is designed to produce an uncorrelated network as $p(\ell_2|\ell_1)$ should not depend on ℓ_1 since edges between pairs of nodes are always independently generated with probability p, and regardless of the degrees of their endpoints. Using this observation, note that the degree of a given node is the binomial random variable $X \in \text{Bin}(n-1,p)$. Hence, $\langle k \rangle \approx \mathbb{E}[X] = (n-1)p$ and $\langle k^2 \rangle \approx \mathbb{E}[X^2] = \text{Var}[X]+\mathbb{E}[X]^2 = (n-1)p(1-p)+((n-1)p)^2 = (n-1)p(1+(n-2)p)$. Therefore, we get that $k_{nn}(\ell) \approx \hat{k}_{nn}(\ell) = \langle k^2 \rangle/\langle k \rangle \approx 1+(n-2)p$. The obtained formula has an intuitive explanation. Take any node v, expose edges adjacent to v, and suppose that its degree is equal to $\ell \geq 1$. Let us concentrate on a random neighbour of v, node u. Node u has v as one of its neighbours and the expected number of its other neighbours is $(n-2)p$.

Note that in the above derivation, in a few places we have used an approximation. Indeed, *after* we take a sample of $\mathcal{G}(n,p)$ we do not expect it to be *exactly* uncorrelated but only approximately, due to the randomness in the graph generation process. However, a sample of $\mathcal{G}(n,p)$ is asymptotically almost surely uncorrelated.

Equation (4.3) reveals a peculiar property of real networks known as the **friendship paradox** first observed by the sociologist Scott L. Feld. He noticed that most people have fewer friends than their friends have, on average. The friendship paradox is an example of how network structure can significantly distort a local observation from an individual node point of view. This observation follows immediately from (4.3) which says that for uncorrelated networks the degree correlation function is *not* equal to $\langle k \rangle$ as one might guess but rather equal to $\langle k^2 \rangle/\langle k \rangle$. Since $\langle k^2 \rangle = \langle k \rangle^2 + \sigma^2$, where σ^2 is the variance of the degrees in the graph, we get that

$$\hat{k}_{nn}(\ell) = \frac{\langle k^2 \rangle}{\langle k \rangle} = \langle k \rangle + \frac{\sigma^2}{\langle k \rangle} \geq \langle k \rangle$$

but it can significantly exceed this trivial lower bound for graphs with nodes of varying degrees (as is typical for many real-world networks). Indeed, for example, for graphs with power law degree distribution with exponent γ and minimum degree δ,

$$\frac{\langle k^2 \rangle}{\langle k \rangle} = \frac{\gamma-2}{\gamma-3}\,\delta = \frac{(\gamma-2)^2}{(\gamma-3)(\gamma-1)}\,\langle k \rangle,$$

provided $\gamma > 3$ (see (2.2) and (2.3)) and can be substantially larger than $\langle k \rangle$ if γ is close to 3. Moreover, if $\gamma \leq 3$, then $\langle k^2 \rangle$ grows with the order of the graph and so the difference is even more pronounced.

In assortative networks, high degree nodes tend to link to other high degree nodes and low degree nodes are more often adjacent to low degree nodes. As a result, for such networks, $k_{nn}(\ell)$ increases with ℓ. On the other hand, in disassortative networks hubs prefer to connect to low degree nodes and vice versa implying that $k_{nn}(\ell)$ decreases as a function of ℓ. There are a few standard ways to proceed from there depending on the assumption on the degree function $k_{nn}(\ell)$.

The first approach assumes that it depends linearly on ℓ and so the **Pearson correlation coefficient** $r \in [-1,1]$ between the degrees of a given edge's endpoints can be used to distinguish between types of correlations. Let (X, Y) be a random variable defined as follows. As it was done in the definition of $p(\ell_1, \ell_2)$ and $p(\ell_2|\ell_1)$, let $e \in E$ be an edge selected uniformly at random from E. Then, a fair coin is tossed to identify the first end of e, node u; the other end is node v. Random variables X and Y represent the degrees of node u and, respectively, node v. Then, the degree correlation coefficient r is simply the **Pearson correlation coefficient** for this pair of random variables—see (1.2). It follows that

$$r = \rho_{X,Y} = \frac{\text{Cov}[X,Y]}{\sqrt{\text{Var}[X]\,\text{Var}[Y]}} = \frac{\mathbb{E}[XY] - \mathbb{E}[X]\,\mathbb{E}[Y]}{\sqrt{\text{Var}[X]\,\text{Var}[Y]}}.$$

In particular,

$$\mathbb{E}[XY] = \frac{1}{m} \sum_{e=uv\in E} \left(\frac{1}{2} \cdot \deg(u) \cdot \deg(v) + \frac{1}{2} \cdot \deg(v) \cdot \deg(u) \right)$$

$$= \frac{1}{m} \sum_{e=uv\in E} \deg(u) \cdot \deg(v) = \sum_{\ell_1,\ell_2\in[\Delta]} \ell_1\ell_2\, p(\ell_1,\ell_2).$$

On the other hand, by symmetry, the means and the variances of X and Y are equal. After a similar computation as above, we get the following useful and convenient formula:

Let $G = (V, E)$ be any unweighted graph on n nodes and $m \geq 1$ edges. The **degree correlation coefficient** is defined as follows:

$$r = \frac{\sum_{\ell_1,\ell_2\in[\Delta]} \ell_1\ell_2\, p(\ell_1,\ell_2) - \left(\sum_{\ell\in[\Delta]} \ell q_\ell\right)^2}{\left(\sum_{\ell\in[\Delta]} \ell^2 q_\ell\right) - \left(\sum_{\ell\in[\Delta]} \ell q_\ell\right)^2}.$$

If $r > 0$, then the network is assortative.
If $r = 0$, then the network is neutral.
If $r < 0$, then the network is disassortative.

After a simple transformation, we get yet another formula for r that is known in the literature:

$$r = \frac{\sum_{\ell_1,\ell_2\in[\Delta]} \ell_1\ell_2\left(p(\ell_1,\ell_2) - q_{\ell_1}q_{\ell_2}\right)}{\left(\sum_{\ell\in[\Delta]} \ell^2 q_\ell\right) - \left(\sum_{\ell\in[\Delta]} \ell q_\ell\right)^2}.$$

However, note that despite the fact that it looks nicer, it is slightly more computationally involved.

The second approach assumes that $k_{nn}(\ell)$ is well approximated as follows: $k_{nn}(\ell) \approx a\ell^\mu$ for some constant $a \in \mathbb{R}_+$ and $\mu \in \mathbb{R}$. This leads us to the next definition.

Let $G = (V, E)$ be any unweighted graph on n nodes and $m \geq 1$ edges. The **correlation exponent** is defined as follows:

$$\mu = \frac{\mathbb{C}\mathrm{ov}\Big(\log(L), \log(k_{nn}(L))\Big)}{\mathbb{V}\mathrm{ar}(\log(L))},$$

where L is a random variable uniformly distributed over the set $\mathcal{D}$.
If $\mu > 0$, then the network is assortative.
If $\mu = 0$, then the network is neutral.
If $\mu < 0$, then the network is disassortative.

In this formulation μ is a coefficient of regression in the model $\log(k_{nn}(\ell)) \approx \log(a) + \mu \log(\ell)$, estimated using the ordinary least squares method. Of course, alternative models could be used instead and they should lead to very similar estimates of μ.

The third approach we would like to mention tries to quantify the tendency for high degree nodes to link together and so to capture the so-called **rich-club behaviour** of a network. In order to do it, one can first compute the number of edges in the graph $G_{\geq\ell} = (V_{\geq\ell}, E_{\geq\ell})$ induced by the set of nodes of degree at least ℓ, that is,

$$\phi(\ell) = |E_{\geq\ell}| = m \sum_{\ell_1 \geq \ell} \sum_{\ell_2 \geq \ell} p(\ell_1, \ell_2),$$

where

$$V_{\geq\ell} = \{v \in V : \deg(v) \geq \ell\} \subseteq V$$
$$E_{\geq\ell} = \{e = uv \in E : u, v \in V_{\geq\ell}\} \subseteq E.$$

In order to benchmark it, we need to compare it with the same quantity evaluated for an uncorrelated graph with the same degree distribution, that is, with

$$\hat{\phi}(\ell) = m \sum_{\ell_1 \geq \ell} \sum_{\ell_2 \geq \ell} \hat{p}(\ell_1, \ell_2) = m \sum_{\ell_1 \geq \ell} \sum_{\ell_2 \geq \ell} q_{\ell_1} q_{\ell_2}$$
$$= \frac{m}{\langle k \rangle^2} \sum_{\ell_1 \geq \ell} \sum_{\ell_2 \geq \ell} (\ell_1 d_{\ell_1})(\ell_2 d_{\ell_2}).$$

After combining the two equations, we get the following definition.

Let $G = (V, E)$ be any unweighted graph on n nodes and $m \geq 1$ edges. The normalized **rich-club coefficient** is defined as follows:

$$\rho(\ell) = \frac{\phi(\ell)}{\hat{\phi}(\ell)} = \langle k \rangle^2 \cdot \frac{\sum_{\ell_1 \geq \ell} \sum_{\ell_2 \geq \ell} p(\ell_1, \ell_2)}{\sum_{\ell_1 \geq \ell} \sum_{\ell_2 \geq \ell} (\ell_1 d_{\ell_1})(\ell_2 d_{\ell_2})}.$$

Values larger than 1 for large values of ℓ and increasing with ℓ indicate the presence of a **rich-club** in a network.

Note that $\rho(1) = 1$ as both $\phi(1)$ and $\hat{\phi}(1)$ are equal to $|E|$. As we already mentioned, if $\rho(\ell) > 1$ for large values of ℓ, then there is an evidence for a rich-club in the network. However, it is also possible that $\rho(\ell) < 1$ which indicates anti rich-club behaviour, that is, large degree nodes tend to avoid other large degree nodes. Finally, let us mention that the degree correlation function $k_{nn}(\ell)$ quantifies local properties of the nodes of degree ℓ whereas the rich-club coefficient $\rho(\ell)$ is a global property of the subgraph induced by large degree nodes. As a result, $\rho(\ell)$ and $k_{nn}(\ell)$ are not correlated in any trivial way; they simply measure different things.

4.4 Structural Cut-offs

In order to decide whether the degree-degree correlation in a given graph is assortative or disassortative, we used the concept of uncorrelated graphs. But is it a fair comparison? This is the main question we would like to address in this section.

First of all, let us note that for a given degree correlation matrix $P = (p(\ell_1, \ell_2))_{\ell_1, \ell_2 \in [\Delta]}$ obtained for some empirical graph G with maximum degree Δ, it is possible to construct a graph $\hat{G}$ that exactly follows the probability distribution specified by $\hat{P} = (\hat{p}(\ell_1, \ell_2))_{\ell_1, \ell_2 \in [\Delta]}$, the counterpart of P for an uncorrelated network. In order to see this, note that $q_\ell = \ell n_\ell / (2m)$ is rational and so $\hat{P}$ has all rational entries; to construct $\hat{G}$ one may, for example, take a union of an appropriately designed family of complete bipartite graphs $K_{i,j}$. However, this potentially might require introducing a graph that is much larger than the one that we plan to benchmark. Hence, it might be considered as comparing apples and oranges. It is natural the matrix P, which describes degree correlations. It is natural to ensure that there exists an uncorrelated graph on the same number of nodes as the original graph G with the same degree distribution d_ℓ as G. In some sense, our benchmark graph should be uncorrelated but still somehow similar to G. As we will see in this section, if we impose this natural restriction on our benchmark graph, then it will not always be possible.

Let us start by noting that if $\ell_1 \neq \ell_2$, then there are $2m \cdot p(\ell_1, \ell_2)$ edges between nodes of degree ℓ_1 and nodes of degree ℓ_2. Similarly, there are $m \cdot p(\ell, \ell)$ edges between nodes of degree ℓ. This means that in an uncorrelated graph, we should expect $2m \cdot \hat{p}(\ell_1, \ell_2)$ edges between nodes of degree ℓ_1 and nodes of degree ℓ_2 if $\ell_1 \neq \ell_2$, and $m \cdot \hat{p}(\ell, \ell)$ edges between nodes of degree ℓ. However, there is a trivial upper bound on the number of edges we can possibly have, since our graph is required to be simple. If $\ell_1 \neq \ell_2$, then we may have up to $n_{\ell_1} n_{\ell_2} = n^2 d_{\ell_1} d_{\ell_2}$ edges, as this corresponds to the number of pairs of nodes of this type. On the other hand, the number of edges between nodes of the same degree is upper bounded by $\binom{n_\ell}{2} = n^2 d_\ell (d_\ell - 1/n)/2$. Combining the two observations together, we get that if one takes two random nodes, one of degree ℓ_1 and another of degree ℓ_2 (ℓ_2 could possibly be equal to ℓ_1), then the expected number of edges between them is equal to

$$\frac{2m\,\hat{p}(\ell_1, \ell_2)}{n^2 d_{\ell_1}(d_{\ell_2} - \delta_{\ell_1 = \ell_2}/n)} \approx \frac{2m\,\hat{p}(\ell_1, \ell_2)}{n^2 d_{\ell_1} d_{\ell_2}} = \frac{\langle k \rangle\,\hat{p}(\ell_1, \ell_2)}{n\,d_{\ell_1} d_{\ell_2}}.$$

In the formula above, δ_A is the **Kronecker delta**: $\delta_A = 1$ if A holds and $\delta_A = 0$ otherwise. But what if the above expectation is larger than one? Since we deal with simple graphs, this is clearly not possible. As a result, comparing graph G to an "imaginary" uncorrelated graph that cannot exist (under the assumption of a given degree distribution d_ℓ and the order of the graph n) is not a fair comparison and should not be used. In such cases, nodes of high degree are "forced" to be adjacent to low degree nodes purely because of the degree distribution of the graph.

A natural question then is whether this problem can occur in practice or whether it is only a theoretical situation that practitioners should not worry about. We immediately see that the star $K_{1,n-1}$ on n nodes creates a problem: the center node is forced to be connected to all the leaves. Additionally, let us investigate some asymptotic behaviour of one of a random graph model to better understand when we might have a problem. Since the networks that we usually deal with are large, while this should help us to build an intuition it should not be used as a decision criterion. Suppose that v and u are both of degree $\ell = o(n)$ and there are $m = \Omega(n)$ edges present in the graph. Using the configuration model (see Section 2.7), the expected number of edges between v and u is equal to

$$\sum_{i=1}^{\ell} \frac{\ell - O(1)}{2m - 2i + 1} \sim \frac{\ell^2}{2m} = \frac{\ell^2}{\langle k \rangle n}.$$

Hence, nodes of degree ℓ that are above the **structural cut-off** threshold

$$k_s = \sqrt{\langle k \rangle\, n}$$

are expected to cause a problem. For example, it is quite common in practice for a graph to follow the power-law degree distribution with exponent $\gamma \in (2,3)$, minimum degree $\delta = \Theta(1)$, and average degree $\langle k \rangle = \Theta(1)$. In such graphs, the maximum degree Δ (the natural cut-off computed in (2.4)) is equal to $\delta n^{1/(\gamma-1)}$ and so is much larger than the structural cut-off k_s introduced above. Of course, as mentioned above, these are only asymptotic observations and one should not only look at the maximum degree to decide whether or not it is justified to use $\hat{k}_{nn}(\ell)$ as a benchmark. We will explain how to do so next.

In order to generate a simple random graph following the prescribed degree distribution $(d_\ell)_{\ell \in \mathbb{N}}$, a theoretically sound approach would be to generate a graph (possibly not simple) using the configuration model (see Section 2.7) and condition it on being simple. However, this requirement is often difficult to satisfy in practice and so a natural algorithm using resampling is very slow and thus not feasible. There are several approaches that are fast enough from a practical point of view while generating a reasonable bias. More details can be found in Section 2.7.

Another common approach is to start with the original graph G that we want to investigate and apply a degree-preserving random process known in the literature as **switching** or **rewiring**. In each step of this process, we select two edges uniformly at random, u_1u_2 and v_1v_2. If after removing these edges we do not create any parallel edges by adding u_1v_1 and u_2v_2, then we accept that switch; otherwise, we ignore it. Note that this operation does not change the degree distribution of the graph. We repeat this step until all edges are switched at least once. Let us note that this algorithm is expected to run for $\Theta(m \ln m)$ steps before it finishes, and it still does not guarantee complete independence (that is, the final outcome of this random process solely depends on the initial graph G we started with). Therefore, in practice, it is common to stop the process prematurely and only perform cm edge switches for some constant c; the larger the value of c, the closer the graph is to the uniform distribution.

Let $\bar{k}_{nn}(\ell)$ be the **randomized degree correlation function** that is computed for the graph obtained at the end of the process mentioned above. If the original function $k_{nn}(\ell)$ and its randomized counterpart $\bar{k}_{nn}(\ell)$ are indistinguishable, then correlations in G are all structural, completely explainable by the degree distribution. In other words, G could be considered an uncorrelated graph. On the other hand, if $\bar{k}_{nn}(\ell)$ does not have a visible correlation with ℓ while $k_{nn}(\ell)$ does, then we can safely apply our previous techniques (use the degree correlation coefficient and/or the correlation exponent) as there seems to be some underlying process that makes the degrees correlated. Finally, it is also possible that $k_{nn}(\ell)$ does not exhibit correlation while $\bar{k}_{nn}(\ell)$ does, which indicates that the process generating G exhibits some level of assortativity or disassortativity despite the fact that $k_{nn}(\ell)$ suggests otherwise. Hence, in short, one should always investigate $k_{nn}(\ell) - \bar{k}_{nn}(\ell)$ as a function of ℓ before drawing any conclusions.

Let us mention that to cross-validate the experiment and check the code, it is quite common to perform another independent switching algorithm in which rewiring steps are always accepted regardless of whether the resulting graph is simple or not. As a result, the resulting graph should have all degree correlations removed, causing the corresponding degree correlation function to be very close to $\hat{k}_{nn}(\ell)$. This approach can serve as an empirical test of whether performing $c\,m$ edge swaps is enough to ensure that the final graph does not strongly depend on G, and adjust constant c if needed.

4.5   Correlations in Directed Graphs

So far we were concerned with undirected graphs but it is straightforward to generalize our tools and observations to directed ones. Let $D = (V, E)$ be any unweighted directed graph on n nodes and $m \geq 1$ directed edges. In order to generalize the degree correlation matrix $P = \big(p(\ell_1, \ell_2)\big)_{\ell_1, \ell_2 \in [\Delta]}$, we may proceed as in the undirected case, namely, we may select one directed edge $uv \in E$ from u to v and investigate its end nodes. However, since each node in a directed graph has an in- and an out-degree, we now have four possibilities to consider. For example, $p^{out,in}(\ell_1, \ell_2)$ is the joint probability that u has out-degree ℓ_1 and v has in-degree ℓ_2. The other three variants are defined similarly and so, for simplicity, we concentrate only on this variant. Note that the degree correlation matrix is symmetric for undirected graphs whereas for directed graphs it can be asymmetric.

The conditional probability $p^{out,in}(\ell_2|\ell_1)$ is the probability that v has in-degree ℓ_2, given that u has out-degree ℓ_1. As for undirected graphs, D is said to be uncorrelated if $p^{out,in}(\ell_2|\ell_1)$ does not depend on ℓ_1. Continuing this analogy, we define the degree correlation function $k_{nn}^{out,in}(\ell)$ to be the average in-degree of the out-neighbours of nodes with out-degree ℓ. The degree correlation coefficient and the correlation exponent are then defined the same way as before. For a rich-club, we may concentrate on the graph induced by the set of nodes of large in-degree, large out-degree, or large total degree.

In summary, in order to completely characterize degree correlations in a directed graph, one needs to consider the four degree functions $(k_{nn}^{in,in}(\ell), k_{nn}^{in,out}(\ell), k_{nn}^{out,in}(\ell), k_{nn}^{out,out}(\ell))$. It might happen that some of them exhibit assortative behaviour whereas others exhibit disassortative behaviour. Hence, the analysis for these graphs is more challenging than the analysis of undirected graphs.

4.6   Implications for Other Graph Parameters

It is good to know whether or not degree correlation is present in the graph. But does it have any important and practical implications that we should be aware of? Let us start the discussion by considering random graphs that usually provide a good intuition on what could potentially happen in practice.

It is well-known that the giant component emerges in binomial random graphs $\mathcal{G}(n, p)$ when $\langle k \rangle = 1$. The same applies to other random graphs generating uncorrelated networks. However, it is also known that for some random graphs generating assortative networks, the phase transition appears earlier, that is, when $\langle k \rangle = x < 1$. We will introduce one such example and do some experiments shortly. The reason behind it is that hubs tend to connect to other hubs which creates a "backbone" that attracts other nodes of the graph. On the other hand, such behaviour is unlikely in disassortative networks and so they form the giant component much later, namely, when $\langle k \rangle = x > 1$. Moreover, these topological differences also affect the order of the giant component.

In order to illustrate this interesting behaviour, let us introduce a simple random graph model that is capable of generating assortative/disassortative networks. The **Xulvi-Brunet–Sokolovs algorithm** generates networks with specified degree correlations. This algorithm is a variant of the switching algorithm that we discussed earlier. This algorithm has one parameter $q \in [0, 1]$ and starts with a simple graph G with the desired degree distribution that can be generated using, for example, the configuration model (see Section 2.7). Then, the algorithm applies the following degree-preserving random process: as before, in each step of the process two edges, $u_1 u_2$ and $v_1 v_2$, are selected uniformly at random from the pairs of edges that span four different nodes. This can be done efficiently with a rejection and resampling strategy. These nodes are then ordered by their degrees. In order to generate an assortative network with probability q, the two edges are rewired in such a way that one edge connects the two nodes with the smaller degrees and the other edge connects the two nodes with the larger degrees; otherwise, the four involved nodes are randomly paired. In order to generate a disassortative network one needs to make one minor change: as before, two random edges on four nodes are selected at random but this time, with probability q, we rewire them such that one edge connects the node of the largest degree with the node with the smallest degree. If the goal is to generate simple graphs, then we always check if a given switching leads to a non-simple graph. If it does, then one simply rejects that switch. By iterating these steps we gradually enhance the network's assortative or disassortative behaviour. In our experiments, the switching is applied until every edge has been successfully rewired at least once but one may stop earlier for a faster, approximated variant of the algorithm.

Figure 4.1 shows results obtained when the initial graph is the binominal random graph $\mathcal{G}(n,p)$. Indeed, the algorithm produces assortative/disassortative networks, provided that $q \neq 0$. For $q = 0$, $k_{nn}(\ell) \approx \langle k^2 \rangle / \langle k \rangle \approx 5$, as expected from the formula derived in Section 4.3. The phase transition for these networks is presented in Figure 4.2. In particular, for $q = 0$ the outcome of the algorithm is simply $\mathcal{G}(n,p)$ and the results are consistent with the asymptotic and theoretical results discussed in Section 2.3.

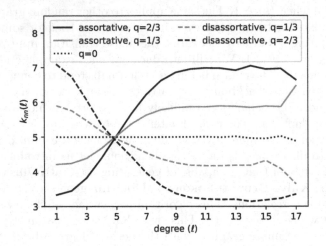

FIGURE 4.1
The degree correlation function $k_{nn}(\ell)$ for the **Xulvi-Burnet–Sokolov's algorithm** applied to $\mathcal{G}(n,p)$ on $n = 2^{16}$ nodes with average degree $\langle k \rangle = 4$. The plots show averages over 64 different initial graphs. For large values of ℓ the plots become unstable because of the small number of nodes of large degree.

These observations can be translated and expressed in the language of percolation theory which describes the behaviour of a network when nodes or edges are randomly removed. This theory has many important applications. For example, suppose that we would like to understand how viruses, news, beliefs, or gossip spread across the contact network of people. Since the fact that two individuals interact with each other does not mean that gossip is exchanged every time they meet, it is natural to consider a random subgraph of the contact network. People that heard the gossip belong to the connected component containing the node that initiated the process. Now, depending on whether the contact network is assortative or disassortative, there might be a drastic difference between the number of people exposed and the number of people affected.

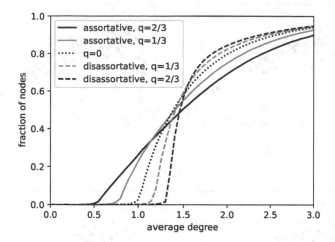

FIGURE 4.2
Order of the giant component as a function of the expected average degree
for the **Xulvi-Burnet–Sokolov's algorithm** applied to $\mathcal{G}(n,p)$ on $n = 2^{16}$
nodes. The plots show averages over 16 different initial graphs.

4.7   Experiments

For experiments, we revisit the US airport graph that we considered in
the previous chapter. We first look at the degree correlation function.
Since the graph is directed, we consider the four possible variants, namely,
$k_{nn}^{in,in}(\ell), k_{nn}^{in,out}(\ell), k_{nn}^{out,in}(\ell), k_{nn}^{out,out}(\ell)$, which we show in Figure 4.3. While
not identical, we see that the four plots are very similar, showing a slightly
negative correlation except for the nodes of small degree. This similarity is
not surprising as in this directed graph the edges very often come in pairs:
when there are flights from A to B, there are very often flights from B to A.

Motivated by this observation, we now consider an undirected, unweighted
version of the airport graph where an edge between two airports is introduced
if there is at least one flight connecting them. In Figure 4.4, we plot the degree
correlation function for this graph, both on a linear and log-log scale. We note
once again the high similarity with the previous plots where edge direction was
considered. In what follows, we retain this undirected graph for the analysis.

Before we move forward, let us note that in Figure 4.4 it is clear that the
relationship between ℓ and $k_{nn}(\ell)$ is non-monotonous and is cap-shaped. This
shows how important it is to visually inspect the data, as just calculating the
degree coefficient r or the correlation exponent μ will not reveal this fact. Still,

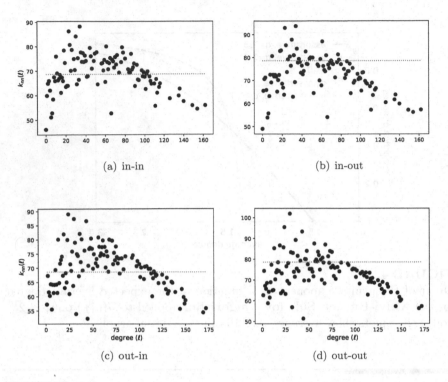

FIGURE 4.3

The four variants of the degree correlation function for the US airport directed graph.

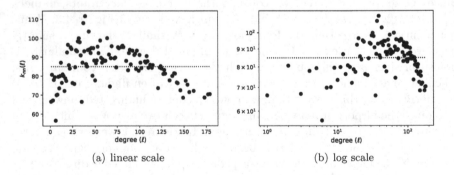

FIGURE 4.4

The degree correlation function for the undirected version of the US airport graph. We show two plots using a linear (a) and a log-log (b) scale.

in the analysis that follows, we use them as they are very convenient if you are analyzing multiple graphs.

Next, we look at the family of induced subgraphs obtained for all states with more than 5 nodes. For each subgraph, we compute the degree correlation coefficient r (often referred to as assortativity) and the correlation exponent μ obtained via a simple linear log-log regression model. In Table 4.5, we show the airports with the most negative or positive assortativity values. Additionally, for the purpose of comparison, we also report the correlation exponents μ. In Figure 4.6, we see a strong correlation between μ and r. In fact, the Pearson correlation values between those two sets of measures is 0.965.

TABLE 4.5
US states subgraphs obtained from the airport graph, sorted by assortativity (r). The corresponding correlation exponents (μ) are included for comparison purpose.

state	nodes	edges	r	μ
NE	6	4	−1.000	−1.000
MN	6	6	−0.833	−0.802
UT	6	6	−0.833	−0.802
AZ	7	9	−0.601	−0.612
MO	11	18	−0.547	−0.345
...	...	...	...	...
SD	6	7	0.263	0.268
ND	6	6	0.400	0.859
AR	7	9	1.000	1.000

Coming back to the family of subgraphs, the first thing we observe is the presence of airports with extreme assortativity values: AR has $r = 1$ and NE has $r = -1$. We plot those small subgraphs in Figure 4.7(a) and (b). Those examples are clear illustrations of the two extreme cases. The AR graph consists of two cliques, so each node has the same degree as its neighbour. On the other hand, in the NE graph every node of degree 1 has a neighbour of degree 2, and nodes of degree 2 have 2 neighbours of degree 1. In Figure 4.7(c) and (d), we show two more examples with slightly larger graphs. In (c), we merge the graphs from North and South Dakota; the resulting graph has assortativity value $r = 0.243$. We notice the presence of a dense region (interconnected high degree nodes) and some tendrils with low degree nodes. In (d), we show the graph for MO which has assortativity value $r = -0.547$. We see that in this graph the low degree nodes mostly connect to the large degree hub nodes.

In order to demonstrate the importance of the structural cut-off, in Figure 4.8(a) we computed assortativity r and the correlation exponent μ for 1,000 random graphs with the same degree distribution as the Dakota

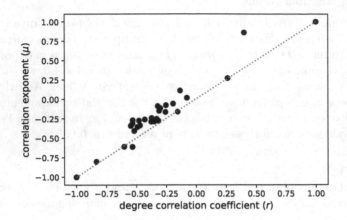

FIGURE 4.6
Assortativity (r) and correlation exponent (μ) for the state subgraphs of the US airport graph.

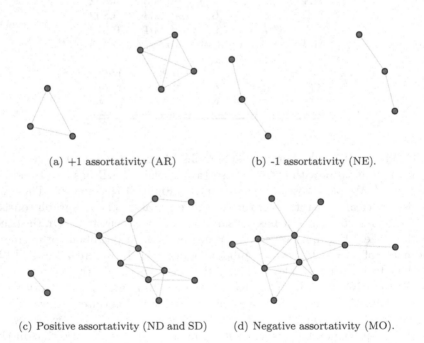

(a) +1 assortativity (AR)

(b) -1 assortativity (NE).

(c) Positive assortativity (ND and SD)

(d) Negative assortativity (MO).

FIGURE 4.7
State subgraphs for the US airport graph with positive ((a) and (c)) and negative ((b) and (d)) assortativity.

(ND+SD) graph generated by the configuration model. The resulting values are quite different from the real graph where $r = 0.243$ and $\mu = 0.383$. However in Figure 4.8(b) we did the same for the Missouri (MO) graph, showing values very similar to the ones for the real graph which are $r = -0.547$ and $\mu = -0.345$. Therefore, in that case, the resulting values for r and μ can mostly be explained by the degree distribution—hubs are forced to be adjacent to small degree nodes. Another way to see this is to plot the degree correlation functions $k_{nn}(\ell)$ for the real graphs and the ones generated by the configuration model—we direct the reader to the companion Python/Julia notebooks available online.

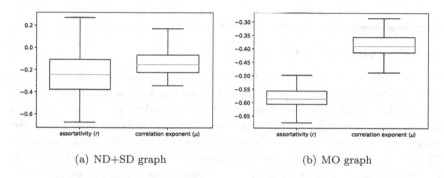

(a) ND+SD graph   (b) MO graph

FIGURE 4.8
Assortativity (r) and correlation exponent (μ) obtained for 1,000 runs of the configuration model using the degree distribution of the airport subgraph for (a) North and South Dakota and (b) Missouri.

We now consider the entire US airport graph, which has an assortativity value of $r = -0.0554$. In Figure 4.9(a), we illustrate the friendship paradox. For each node, we compare its degree with the average degree of its neighbours. We draw a line with unit slope to highlight the fact that the region above that line is much denser. This is due to the fact that there are very many low degree nodes (smaller airports) which mostly tend to connect to hub airports, which explains the presence of this "paradox." In Figure 4.9(b), we compute the rich-club ratio $\rho(\ell)$ for all values of ℓ. We see that the curve starts at $\rho(1) = 1$, increases slightly before decreasing gradually. We conclude that there is no indication of a rich-club phenomenon here.

In Figure 4.10, we re-visit the GitHub developer graph and the European electric grid network introduced in Chapter 1, comparing the degree correlation functions. We see that not only are the degree distributions quite different but also that the type of connections and the assortativity values are very different. In the GitHub graph, low degree nodes are mostly connected to high degree hub nodes, and this effect drops very quickly as we consider higher degree nodes. For the grid network on the other hand, the behaviour is the opposite despite much more homogeneous values for $k_{nn}(\ell)$.

Finally, in Figure 4.11, we look again at the rich-club coefficient but this time for a well-known graph that exhibits a rich-club behaviour, first studied by Watts and Strogatz in 1998 and available here[1]. The nodes are movie actors and there is an edge between two actors if they co-appear in at least one movie. In that plot, we clearly observe the rich-club phenomenon. This can be explained by the fact that famous actors tend to play in many movies (high degree) and with other famous actors (rich-club phenomenon).

4.8   Practitioner's Corner

Beyond being informative about the graph's degree distribution, the degree correlation function provides information about the local structure of a graph. It can be computed for undirected as well as directed graphs, where there are four different variants due to combinations of in- and out- degrees. The degree correlation coefficient r is used to measure whether nodes tend to connect with other nodes of similar degree (positive) or not (negative); this can also be estimated via the correlation exponent μ obtained via regression on the degree correlation function. By considering random models such as the configuration model or switching model, we can validate whether the assortativity suggested by some measures is truly present in a graph, or if it is just an artefact of its degree distribution. This is related to the concept of a structural cut-off which indicates that high degree nodes can expect to have more than one edge between them in an uncorrelated counterpart, which is problematic for

---

[1] www.complex-networks.net

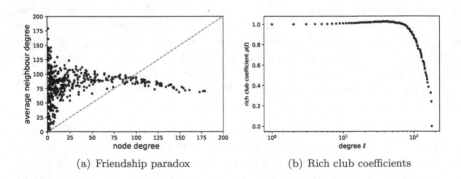

(a) Friendship paradox   (b) Rich club coefficients

FIGURE 4.9
The friendship paradox (a) and rich-club behaviour (b) for the US airport graph.

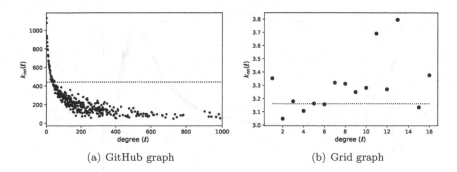

(a) GitHub graph   (b) Grid graph

FIGURE 4.10
Degree correlation function for (a) the GitHub developers graph (with $r = -0.075$) and (b) the Europe electric grid network (with $r = 0.014$).

simple graphs. Finally, another structure that is sometimes seen in graphs is the presence of a "rich-club" phenomenon, where high degree nodes have lots of edges between them. This can be tested with the rich-club coefficient.

4.9   Problems

In this section, we present a collection of potential practical problems for the reader to attempt.

1. For each of the three GitHub Developers graphs (the ml developers, the web developers, and the original one), do the following:

   a. plot the degree correlation function $k_{nn}(\ell)$ and its uncorrelated counterpart $\hat{k}_{nn}(\ell)$;

   b. find the degree correlation coefficient r,

   c. find the correlation exponent μ,

   d. plot the rich-club coefficient $\rho(\ell)$.

2. Starting with the original GitHub Developers graph, apply the switching method to get the randomized degree correlation function $\bar{k}_{nn}(\ell)$. Compare it with the degree correlation function $k_{nn}(\ell)$ and its uncorrelated counterpart $\hat{k}_{nn}(\ell)$. What is your conclusion?

3. Starting with the original GitHub Developers graph, apply the switching method but instead of waiting for all edges to be switched at least once, perform only cm edge switchings for $c \in \{0.5, 1, 2\}$

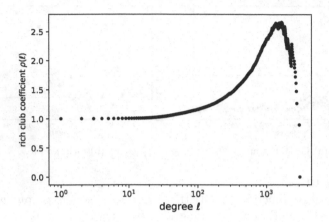

FIGURE 4.11
The rich-club phenomenon is clearly seen in a graph of co-appearance of movie actors.

(m is the number of edges). For each experiment, plot the (approximated) randomized degree correlation function $\bar{k}_{nn}(\ell)$. Independently check how many edges have *not* been switched. (After cm edge switchings, we randomly switch $2cm$ edges and theoretical results imply that we expect e^{-2c} fraction of edges to remain not switched. You may check if your experiment returns a similar answer.)

4. Perform a similar experiment on 5 graphs obtained by applying **Xulvi-Burnet–Sokolovs** algorithm to $\mathcal{G}(n,p)$. (The algorithm is quite slow but it should be quick with, say, $n = 2^8$ and 10 repeats.) Sample pairs of nodes to estimate the average distance between two nodes in the graph instead of investigating the fraction of nodes in the giant component. What is your conclusion?

5. Use the **Chung-Lu** model with $n = 10{,}000$ nodes to generate three graphs with power law degree distribution ($\gamma = 2.1$, $\gamma = 2.5$, and $\gamma = 2.9$), $\delta = 1$, and $\Delta = 100$—see Figure 2.6. For each of them, investigate the friendship paradox and rich-club behaviour similarly as we have done in Figure 4.9.

6. Generate power-law degree distribution **d** with $n = 10{,}000$, $\gamma = 2.1$, $\delta = 1$, and $\Delta = 500$.

   a. Generate the random graph $\mathcal{P}_{n,\mathbf{d}}$ with this degree sequence (it is implemented in Python's `iGraph`—use option `simple` in function `Graph.Degree_Sequence`). Plot the degree correlation function $k_{nn}(\ell)$ for this graph. Since this graph is not necessarily simple, $k_{nn}(\ell)$ should be close to a straight line $\hat{k}_{nn}(\ell) = \langle k^2 \rangle / \langle k \rangle$ which you should plot too.

b. Generate a random graph with this degree sequence insisting that the outcome is a simple graph (it is implemented in Python's iGraph—for example, use option vl in function Graph.Degree_Sequence that uses the implementation of **Viger and Latapy** that creates a connected graph). Plot the degree correlation function $k_{nn}(\ell)$ for this graph. Since this graph is simple, $k_{nn}(\ell)$ should be close to $\bar{k}_{nn}(\ell)$, the randomized degree correlation function.

What is the conclusion?

4.10   Recommended Supplementary Reading

The Internet was the first complex network in which degree-degree correlations were explored and found:

- R. Pastor-Satorras, A. Vazquez, A. Vespignani. "Dynamical and correlation properties of the Internet". *Phys. Rev. Lett.* **87** (2001), 258701. (Degree Correlation Function)

The movie actor dataset introduced in this chapter was first studied in:

- D.J. Watts and S.H. Strogatz, "Collective dynamics of 'small-world' networks", Nature 393 (1998), 440–442.

There are a few original papers introducing the corresponding concepts and one survey:

- M.E.J. Newman. "Assortative mixing in networks". *Phys. Rev. Lett.* **89** (2002), 208701. (Degree Correlation Coefficient)

- S.L. Feld. "Why your friends have more friends than you do", *American Journal of Sociology* **96(6)** (1991), 1464–1477. (Friendship Paradox)

- V. Colizza et al. "Detecting rich-club ordering in complex networks". *Nat. Phys.* **2** (2006), 110–115. (Rich Club)

- M. Boguñá, R. Pastor-Satorras, A. Vespignani. "Cut-offs and finite size effects in scale-free networks". *Eur. Phys. J. B* **38** (2004), 205–209. (Structural Cut-off)

- N.C. Wormald. "Models of random regular graphs." *Surveys in Combinatorics.* 239–298, London Math. Soc. Lecture Note Ser., 267, Cambridge Univ. Press, Cambridge, 1999. (Switching)

- R. Xulvi-Brunet, I.M. Sokolov. "Changing correlations in networks: assortativity and dissortativity". *Acta Phys. Pol. B* **36** (2005), 1431–1455. (Xulvi-Brunet–Sokolovs algorithm)

5

Community Detection

5.1 Introduction

A network has community structure if its set of nodes can be split into a number of subsets such that each subset is densely internally connected. For example, many social networks consist of communities based on the common location of their users, their interests, occupation, gender, age, etc. Dense clusters of connected neurons in the brain are often synchronized by their firing patterns. In the web graph, nodes that belong to the same cluster correspond to web pages on a similar topic. In protein-protein interaction networks, proteins that belong to the same community are often associated with a particular biological function within the organism.

There are a lot of partitions of a given set of n nodes, even if n is very small. Hence, finding a partition that accurately represents the community structure of a given network is a challenging but important problem for a number of reasons. Communities allow us to look at the network from a large distance and see the big picture by creating a large scale map of a network with individual communities represented as meta-nodes. They help us better understand the function of the system represented by the network. They also let us classify the nodes based on the position they have in their own clusters and how they are connected to other clusters. As a result, we may better understand their roles and importance.

This chapter is structured as follows. We first introduce basic definitions and properties that are expected to be present in a partition returned by a good community detection algorithm (Section 5.2). The problem we try to address in this chapter is unsupervised in nature and so we typically do not know how to benchmark a given algorithm run on a particular real-world network. As a result, there is a need to be able to construct synthetic models with a given community structure and to then rigorously evaluate them (Section 5.3). Since finding communities is an important task, it has generated many multidisciplinary research projects. As a result, algorithms use different ideas and approaches, such as the graph modularity function (Section 5.4), hierarchical clustering (Section 5.5), label propagation, spectral bisection method, and the information-theoretic method (Section 5.6). As usual, we finish the

DOI: 10.1201/9781003218869-5

chapter with experiments (Section 5.7) and provide some tips for practitioners (Section 5.8).

Finally, let us mention that in this chapter we concentrate on non-overlapping communities. Overlapping communities are separately discussed in Chapter 8.

5.2   Basic Properties of Communities

Most of us have a good intuition of what it means to be a part of some community. For example, one might explicitly be a member of a local chess club (by paying an annual membership fee or by simply attending their weekly meetings) but there are many more implicit communities one might belong to, often involving networks. Indeed, one might belong to the community consisting of mathematicians within the collaboration network, regardless of whether that person has an academic job, higher education, etc. The important question is then what makes that person a part of it and how one may detect such memberships. There are some properties that will allow us to formally define communities in networks and then to design algorithms to find them but let us stress the fact that such properties are simply a natural implication of an underlying structure (that might or might not be visible to us). For example, the community of mathematicians publishing papers together consists of people that are interested in mathematics in the first place and, as a result, they naturally start collaborating with each other. Such behaviour and interactions between them creates a graph with certain properties. Our goal is to use these properties to uncover communities and often we have no other information but the graph. This approach can be viewed as the *reversed engineering process* that tries to uncover a hidden, underlying attributes of nodes, the so-called ground truth, based on a graph that is visible to us.

Ground Truth

In order to explain the concept behind the **ground truth**, let us consider a social network that has received a lot of attention in the context of community detection. The *Zachary's Karate Club Network* is a small graph, available within `igraph`, that represents social interactions between 34 members of a karate club. Since the club was small, members of the club knew each other well but not all of them interacted outside of the club. Such interactions can be represented by the 78 edges in the corresponding graph that we will frequently analyze in this book. A conflict between the club's president and the instructor split the members of the club into two groups which suddenly revealed the underlying community structure, the ground truth.

In general, ground truth can be associated with some attributes assigned to the nodes that influence how edges in a graph are formed. In the example of Zachary's Karate Club, the ground truth is simply a partition of nodes of the associated graph into two groups but, in general, the number of parts can be arbitrary.

Let V be any finite set (for example, V could be the set of nodes of a graph G). A **partition** of V is a grouping of its elements into non-empty subsets, in such a way that every element is included in exactly one subset. In other words, $V_1, V_2, \ldots, V_\ell$ is a partition if and only if $V_1 \cup V_2 \cup \ldots \cup V_\ell = V$ and $V_i \cap V_j = \emptyset$ for any $1 \le i < j \le \ell$.

In this chapter, we mostly focus on problems where it is assumed that the ground truth is a partition, informing us about the structure of communities in the graph. However, in general, other ground truths may be considered. The simplest extension of partitioning nodes of a graph is to allow each node to be a member of more than one community or be part of no community. As mentioned in the introduction, overlapping communities are separately discussed in Chapter 8.

The fact that we have access to the ground truth explaining an underlying structure is a very rare situation. As a result, many algorithms are benchmarked using this small graph as an initial test. Of course, one should not draw any serious conclusions based on such a small example. In order to test the performance of various algorithms, one should instead use some large but artificially generated graphs with community structures. We will discuss a few of such benchmark networks in Section 5.3. There are also some performance measures such as graph modularity that are unsupervised in nature and so they do not require the ground truth for an evaluation procedure. We will introduce them in Section 5.4.

Definition

Let $G = (V, E)$ be a graph with some community structure. Each **community** in G corresponds to a subset C of V, the set of nodes, that induces a subgraph $G[C]$ (see Section 1.9 for a formal definition). It is often expected that $G[C]$ is connected but, more importantly, nodes in a community are more likely to be connected to other members of the same community than to nodes in other communities. This leads us to the following natural definitions that can be applied to both weighted and unweighted graphs.

Let $G = (V, E)$ be any graph, and let $C \subseteq V$. The **internal degree** $\deg^{int}(v)$ (with respect to C) of node v is the number of neighbours of v that belong to C, that is,

$$\deg^{int}(v) = |N(v) \cap C|.$$

The **external degree** $\deg^{ext}(v)$ (with respect to C) of node v is the number of neighbours of v outside of C, that is,

$$\deg^{ext}(v) = |N(v) \setminus C| = \deg(v) - \deg^{int}(v).$$

The following two concepts are natural and common in the literature.

Let $G = (V, E)$ be any graph, and let $C \subseteq V$. Set C forms a **strong community** if each node in C has more neighbours in C than outside of C, that is, for each $v \in C$ we have

$$\deg^{int}(v) > \deg^{ext}(v). \tag{5.1}$$

One may relax this notion by considering the average degree inside the community C (over all nodes in C) and compare it with the total average degree, instead of insisting that all nodes in C satisfy the desired inequality.

Let $G = (V, E)$ be any graph, and let $C \subseteq V$. Set C forms a **weak community** if

$$\sum_{v \in C} \deg^{int}(v) > \sum_{v \in C} \deg^{ext}(v). \tag{5.2}$$

Of course, if C satisfies (5.1), then it also satisfies (5.2) so each strong community is a weak community but some weak communities might not be strong.

There are many other notions that one may use to formally define communities. Consider, for example, the following scenario. Suppose that two researchers have the same number of friends on Instagram (say, 100) but they belong to two different communities. The first researcher, Alice, is part of a large community (say, she is a data scientist) whereas the second researcher, Bob, belongs to a small community (say, he is a mathematician doing some esoteric part of mathematics). Suppose that 60% of Instagram friends of Alice do data science making her a clear member of that community. What about Bob? Should we expect that more than 50% of his friends to be working in his field of research? We believe the answer is no. In fact, it might be the case that there are less than 50 people around the world working on this subject! It seems that it makes more sense for the number of internal neighbours of a given node to be proportional to the size of the community this node belongs to. As long as the probability that a given node is adjacent to another member of its community is larger than the probability of being adjacent to a random node in the whole graph, this node is a legitimate member of this community. This notion is strongly related to the concept of modularity that we will discuss in Section 5.4 and motivates our last definition.

Let $G = (V, E)$ be any graph, and let $C \subseteq V$ ($C \neq \emptyset$ and $C \neq V$). C is a **community** if each node in C has proportionally more neighbours in C than outside of C, that is, for each $v \in C$ we have

$$\frac{\deg^{int}(v)}{|C|} > \frac{\deg^{ext}(v)}{|V \setminus C|}. \tag{5.3}$$

Node Roles

Let $G = (V, E)$ be a graph on the set of nodes V. Suppose that we are given a partition $\mathcal{A} = \{A_1, A_2, \ldots, A_\ell\}$ of V; $\mathcal{A}$ could be the partition representing the ground truth or a partition identified by some community detection algorithm. In this sub-section, we will use $\deg_{A_i}(v)$ to denote the number of neighbours of v in part A_i, that is, $\deg_{A_i}(v) = |N(v) \cap A_i|$. We will also continue using the definitions from the previous sub-section, in particular, the internal degree of node v is denoted by $\deg^{int}(v) = \deg_{A_i}(v)$ for the unique $i \in [\ell]$ for which $v \in A_i$.

Our goal is to try to quantify the role played by each node within a network that exhibits community structure. The first definition captures how strongly a particular node is connected to other nodes within its own community, completely ignoring edges between communities.

Let $G = (V, E)$ be any graph, and let $\mathcal{A} = \{A_1, A_2, \ldots, A_\ell\}$ be a partition of V. The **normalized within-module degree** of a node v (with respect to $\mathcal{A}$) is defined as follows:

$$z(v) = \frac{\deg^{int}(v) - \mu(v)}{\sigma(v)},$$

where $\mu(v)$ and $\sigma(v)$ are respectively the mean and the standard deviation of $\deg^{int}(u)$ over all nodes in the part v belongs to.

Note that in the definition above we assumed that the graph induced by the part node v belongs to is *not* regular (that is, $\sigma(v) \neq 0$). Note also that $z(v)$ is the familiar Z-score as it measures how many standard deviations the internal degree of v deviates from the mean. If node v is tightly connected to other nodes within the community, then $z(v)$ is large and positive. On the other hand, $|z(v)|$ is large and $z(v)$ is negative when v is loosely connected to other peers.

Another important aspect is to capture how edges (both internal and external) are distributed between all parts of the partition $\mathcal{A}$.

Let $G = (V, E)$ be any graph, and let $\mathcal{A} = \{A_1, A_2, \ldots, A_\ell\}$ be a partition of V. The **participation coefficient** of a node v (with respect to $\mathcal{A}$) is defined as follows:

$$p(v) = 1 - \sum_{i=1}^{\ell} \left(\frac{\deg_{A_i}(v)}{\deg(v)} \right)^2.$$

The participation coefficient $p(v)$ is equal to zero if v has neighbours exclusively in one part (most likely in its own community). In the other extreme situation, the neighbours of v are homogeneously distributed among all parts and so $p(v)$ is close to the trivial upper bound of

$$1 - \sum_{i=1}^{\ell} \left(\frac{\deg(v)/\ell}{\deg(v)} \right)^2 = 1 - \frac{1}{\ell} \approx 1.$$

This coefficient is a natural way to measure concentration. Indeed, in economics, $1 - p(v)$ is called the **Herfindahl–Hirschman index** and is essentially equivalent to the **Simpson diversity index** used in ecology, the **inverse participation ratio (IPR)** in physics, and the **effective number of parties index** in politics.

The two definitions we just introduced can help us to assign roles to all nodes in the network based on their topological function within the communities. The following classification has been proposed in the literature. If $z(v) > 2.5$, then node v is classified as a **hub**; otherwise, it is a **non-hub**. These two classes are then refined by investigating the participation coefficient. Non-hubs are partitioned into the following four classes:

- **ultra-peripheral nodes** (almost all edges are internal) if $p(v) < 0.05$,

- **peripheral nodes** (most edges are internal) if $0.05 \leq p(v) < 0.62$,

- **connector nodes** (most edges are external) if $0.62 \leq p(v) < 0.80$,

- **kinless nodes** (neighbours are homogeneously distributed) if $p(v) \geq 0.80$.

On the other hand, hubs are partitioned into three classes:

- **provincial hubs** if $p(v) < 0.30$,

- **connector hubs** if $0.30 \leq p(v) < 0.75$,

- **kinless hubs** if $p(v) \geq 0.75$.

The above classification might look rather arbitrary but it certainly provides some useful insight about the roles particular nodes have. For example, a provincial hub has a pivotal role within its own community whereas a connector hub is important for transferring information between communities.

The Number of Partitions

In order to find communities in a given graph $G = (V, E)$, we need to partition the set of nodes V into an arbitrary number of parts. The **Bell number** B_n counts the number of possible partitions of a set of size n. Starting with $B_0 = B_1 = 1$, the first few Bell numbers are: 1, 1, 2, 5, 15, 52, 203, 877, 4140. This sequence looks innocent but the rate of growth of the first few terms is misleading; the number of partitions grows very fast as a function of n. Indeed, trivially, the number of partitions into 2 parts is equal to 2^n, already exponentially many, and so we get that $B_n \geq 2^n$. In fact, B_n grows much faster than that.

The Bell numbers can be computed using the following recurrence relation involving binomial coefficients: for any $n \in \mathbb{N}$,

$$B_{n+1} = \sum_{k=0}^{n} \binom{n}{k} B_k.$$

Indeed, in order to partition a set of $n + 1$ elements, one may first remove a part containing the first element and then take any partition of the remaining elements. There are $\binom{n}{k}$ choices for the k items that remain after one part is removed and B_k choices of how to partition them. Of course, there are many other interesting relations between Bell numbers. Similarly, several asymptotic formulas for the Bell numbers are known, including the following:

$$\frac{\ln B_n}{n} = \ln n - \ln \ln n - 1 + o(1),$$

which implies that

$$B_n = \left(\frac{n}{(e + o(1)) \ln n} \right)^n.$$

Hence, indeed, B_n grows much faster than any exponential function.

These observations have an important implication for our problem. Even for very small graphs on, say, 20 nodes we will not be able to investigate all partitions. Since we cannot do it for such small graphs, there is no hope for larger ones. However, our ultimate goal is to find communities in enormous graphs such as the collaboration network or the Facebook graph. Hence, all algorithms we investigate in this chapter are heuristic in nature. Our hope is to find a decent partition that is good enough for the application at hand; we should never expect to find the best one.

5.3   Synthetic Models with Community Structure

In order to compare clustering algorithms, one can use a network with an explicitly given ground truth such as the Zachary's Karate Club network mentioned above. However, this small network should be treated as a toy-example rather than as a serious benchmark. Unfortunately, there are not too many other networks available that have a ground truth available to be used. Alternatively, one can use some other quality measure, for example, the modularity function mentioned in Section 5.4. Indeed, modularity is not only a global criterion to define communities and a way to measure the presence of community structure in a network but, at the same time, it is often used as a quality function of community detection algorithms. However, it is not a fair benchmark, especially for comparing algorithms such as **Louvain** and **Ensemble Clustering** (see Section 5.4) that find communities by trying to optimize the very same modularity function!

In order to evaluate algorithms in a fair and rigorous way, one should compare algorithm solutions to a large synthetic network with an engineered ground truth. Let us introduce a few random graph models with community structure before we show how to use them to make such evaluations. Of course, there are many more models of complex networks so we only scratch the surface by presenting three models that are, arguably, the most related to community detection. Other models worth mentioning are **BTER** and **Re-CoN**.

Stochastic Block Model

We start with the **stochastic block model**, which is a simple and natural generalization of the **binomial random graph** model $\mathcal{G}(n,p)$ that we introduced in Section 2.3. This model produces random graphs consisting of communities but edges are still generated independently, making this model relatively easy to study from a theoretical point of view. It is also easy to generate these graphs in practice.

Let $\mathbf{P} = (p(i,j))_{i,j\in[\ell]}$ be a symmetric $\ell \times \ell$ matrix whose entries are in $[0,1]$ and let $\mathbf{N} = (n_i)_{i\in[\ell]}$ be a vector of ℓ natural numbers. The **stochastic block model** $\mathcal{S}(\mathbf{P},\mathbf{N})$ can be generated by starting with an empty graph on the set of nodes $[n] = \{1,2,\ldots,n\}$ with $n = \sum_{i\in[\ell]} n_i$ partitioned into ℓ subsets $C_1, C_2, \ldots, C_\ell$, called **communities**, with $|C_i| = n_i$, $i \in [\ell]$. For each pair of nodes $u,v \in [n]$ with $u < v$, $u \in C_i$, and $v \in C_j$, we independently introduce an edge uv in $\mathcal{S}(\mathbf{P},\mathbf{N})$ with probability $p(i,j)$.

The **planted partition model** is a variant of the **stochastic block model** in which the values of the probability matrix $\mathbf{P}$ has values of p on the diagonal and values of q for non-diagonal entries. As a result, two nodes within the same community share an edge with probability p whereas two nodes in different communities share an edge with probability q. The case where $p > q$ is called an **assortative** model, while the case of $p < q$ is called a **disassortative** model. (Recall that such terms were already used in Section 4.2 in the context of degree correlations but there is no relationship between the two notions.) Finally, if the probability matrix is a constant matrix (that is, $p(i,j) = p$ for all $1 \leq i \leq j \leq \ell$), then the **stochastic block model** recovers the classical **binomial random graph** $\mathcal{G}(n,p)$.

Lancichinetti–Fortunato–Radicchi (LFR) Model

The **stochastic block model** $\mathcal{S}(\mathbf{P}, \mathbf{N})$ introduced above generates a random graph in which all nodes that belong to the same community have exactly the same distributions of their degrees, that is, each of them has a degree that is a sum of the same binomial distributions (one corresponding to their internal degree and one to their external one). In particular, they have the same expected degree. On the other hand, the degree distributions as well as the community sizes of most real networks follow power-law distributions (see Section 2.4). The community sizes in $\mathcal{S}(\mathbf{P}, \mathbf{N})$ are provided as an input vector $\mathbf{N}$ but it is left for the user to make sure they follow some realistic distribution.

The next random graph model, the **LFR** (**Lancichinetti–Fortunato–Radicchi**) benchmark, allows for heterogeneity in the distributions of both nodes degrees and of community sizes. As a result, it has become a standard and extensively used method for generating artificial networks with communities. Since the formal definition and implementation details are quite involved, we first define (rather vaguely) the **LFR** model before we describe a few specific ingredients of the algorithm that generates it.

> Let $\gamma \in [2,3]$, $\tau \in [1,2]$, and $\mu \in [0,1]$ (μ is called the **mixing parameter**). The **LFR** model $\mathcal{L}(\gamma, \tau, \mu)$ generates random graphs with community sizes and node degrees following truncated power law distributions with exponents τ and γ respectively. Each node aims to have a μ fraction of its neighbours outside of its own community.

The process of generating **LFR** graphs follows the following steps.

- Randomly generate the degree distribution following a truncated power law distribution with exponent γ and the desired average degree $\langle k \rangle$. In particular, based on the generated degree sequence $\mathbf{d} = (d_1, d_2, \ldots, d_n)$, the minimum and the maximum degree (δ and, respectively, Δ) are determined.

- Use the **random graph with degree sequence d** (see Section 2.7) to create a graph with the desired degree sequence $\mathbf{d}$.

- Randomly generate the community sizes following a power law distribution with exponent τ. The sum of all sizes must be equal to n, the desired number of nodes. The minimum and the maximum community sizes (n_{min} and, respectively, n_{max}) must satisfy some properties so that every non-isolated node belongs to one community. In particular, it is required that $n_{min} > \delta$ and $n_{max} > \Delta$.

- Initially, no node is assigned to any community. Recall that the goal for each node is to have a fraction of $(1 - \mu)$ of its neighbours within the same community. In order to achieve this, each node is randomly assigned to a community. If the desired number of neighbours of a given node within the community does not exceed the community size, then the node is added to the community; otherwise, it stays unassigned. In the following iterations, each unassigned node is randomly assigned to some community. If that community is complete (that is, its desired size is already reached), then a randomly selected node from that community becomes unassigned. The process is repeated until all the communities are complete, that is, all the nodes are assigned to a community.

- In order to achieve the desired property for the fraction of internal neighbours expressed by the mixing parameter μ, several random rewiring of edges are performed, where the preference is to rewire edges that do not link nodes which have common neighbours. Each rewiring preserves the degree distribution but aims to improve the ratio between the number of internal and the number of external neighbours. The process stops when good approximation is achieved for all nodes.

Artificial Benchmark for Community Detection (ABCD)

The **LFR** benchmark mentioned above is a very good model and has become a standard benchmark for experimental studies, both for disjoint and for overlapping communities. However, in order to provide motivation for our next benchmark graph, let us point out some potential issues.

Scalability: In order to generate a random graph following a given degree sequence and the desired ratio between internal and external neighbours, the **LFR** benchmark uses the fixed degree sequence model and then the edge switching Markov chain algorithm to obtain the desired community structure once the stationary distribution is reached. The convergence process is inherently slow and so the model has clear scalability limitations that are known to both academics and practitioners.

Many Variants and Lack of Theoretical Foundations: Since the most computationally expensive part of the **LFR** benchmark is edge switching, many fast implementations stop the process before the stationary distribution is reached. There are at least two negative implications of this situation. First of all, there are many variants of this benchmark model and one can

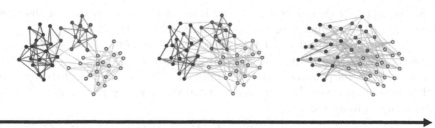

FIGURE 5.1
Examples of graphs generated by the **LFR** algorithm. All graphs have the same degree distribution and community sizes but have different values of the mixing parameter: $\mu \in \{0.1, 0.3, 0.95\}$.

only create "**LFR**-type" graphs and graphs generated by different implementations can have slightly different properties. This is certainly not expected from benchmark graphs that should provide a rigorous, fair, and reproducible comparison.

The lack of a simple and clear description of the algorithm has another negative aspect, namely, it is challenging to analyze the **LFR** model theoretically. It is unfortunate, as more theoretical research on models with community structure might shed some light on how communities are formed and help us design better and faster clustering algorithms.

Communities are Unnaturally-defined: The mixing parameter μ, the main parameter of the **LFR** model guiding the strength of the communities, has a non-obvious interpretation and so can lead to unnaturally-defined networks. (See Figure 5.1 for an illustration.) Indeed, the model aims to keep the fraction of inter-community edges at approximately μ. In one of the two extremes, when $\mu = 0$, all edges are within communities. On the other hand, when $\mu = 1$, **LFR** generates pure "anti-communities" with no edge present in any of the communities. This is clearly an undesirable property that leads to unnaturally-defined communities. The threshold value of μ that produces pure random graphs that are community agnostic is "hidden" somewhere in the interval $[0, 1]$. It is possible to compute this threshold value but the formula is quite involved and not widely known.

Densities of Communities: The **LFR** model aims to generate a graph in which a fraction of $1 - \mu$ edges adjacent to a given node stay within the community of that node. This property is required to hold for all nodes, regardless of the size of the community that they belong to. As a result, small communities will become much denser than large communities. It is not clear that this property is desirable, especially in the case of unbalanced community sizes which the model is aiming for. Indeed, it seems that larger clusters should capture a proportionally larger fraction of edges—recall the discussion right after the definition of **community** (see Equation (5.3)).

The approach used in **LFR** (which we call a **local** variant) seems to be inherited from the definition of **strong community** (see Equation (5.1)). The next model proposes another approach (that we call a **global** variant) that builds on top of the definition of **community** (see Equation (5.3)) that, arguably, is more natural.

Now, we are ready to introduce the Artificial **B**enchmark for **C**ommunity **D**etection (**ABCD**). As we did with **LFR**, we provide a "bare minimum" definition (that is almost the same as for **LFR**) before we describe a few details of the implementation.

> Let $\gamma \in [2,3]$, $\tau \in [1,2]$, and $\xi \in [0,1]$ (ξ is called the **mixing parameter**). The **ABCD** model $\mathcal{A}(\gamma, \tau, \xi)$ generates random graphs with community sizes and node degrees following truncated power law distributions with exponents τ and γ respectively. The mixing parameter ξ controls the fraction of edges that are between communities.

The process of generating **ABCD** graphs follows the following steps.

- As we did in **LFR**, we start by randomly generating the degree distribution $\mathbf{d}$ following a truncated power law distribution with exponent γ and the community sizes following a power law distribution with exponent τ. (Alternatively, they may be provided as deterministic parameters of the model). In particular, ℓ is the number of communities.

- The model can be viewed as a union of $\ell+1$ independent random graphs G_i ($i \in [\ell] \cup \{0\}$)—one for each community, and one for the whole graph. As a result, one can view it as a generalization of the **double round exposure method** (also known in the literature as "**sprinkling**"). We start with the background graph G_0 and "sprinkle" additional edges within communities that come from graphs G_i ($i \in [\ell]$); the smaller the value of ξ, the stronger the ties between members of the same cluster are.

- Parameter $\xi \in [0,1]$ controls the fraction of edges that are between communities. We split the degree sequence $\mathbf{d}$ into $\mathbf{y}$ and $\mathbf{z}$ as follows, keeping the same value of ξ for each node ($\mathbf{y}$ will be associated with communities and $\mathbf{z}$ will be associated with the background graph):

$$\mathbf{y} = (y_1, \ldots, y_n) = (1-\xi)\mathbf{d} = ((1-\xi) \cdot d_1, \ldots, (1-\xi) \cdot d_n),$$
$$\mathbf{z} = (z_1, \ldots, z_n) = \xi\mathbf{d} = (\xi \cdot d_1, \ldots, \xi \cdot d_n).$$

This splitting strategy corresponds to the more natural, global, variant. In order to generate graphs that are even closer to the original **LFR** model, one may alternatively use another strategy for splitting, the local variant. Both are easily available to the user.

- The process of assigning nodes into communities is quite involved so we will not explain it here. However, let us mention that it efficiently selects uniformly at random one admissible assignment which, in particular, guarantees that a simple graph satisfying all the desired properties can be constructed.

- Since we aim to generate a graph which follows the degree sequence **d**, the two vectors (**y** and **z**) need to be adjusted so that not only are they integer-valued but the corresponding subsequences (for each cluster and the background graph) are graphic.

- We use the **random graph with fixed degree sequence** (see Section 2.7) to independently generate the **background graph** G_0 and the **cluster graphs** G_i (for $i \in [\ell]$). The final multi-graph $\mathcal{A}(\gamma, \tau, \xi)$ is defined as the union of graphs G_i ($i \in [\ell] \cup \{0\}$).

For theoretical results on **ABCD**, one may simply stay with multi-graphs that might not be simple but the expected number of loops and parallel edges is small, especially for sparse graphs, so that most results should not be affected by their presence. From practical point of view, if one insists on the final graph to be simple, then the following additional step can be performed.

- For each parallel edge uv (or a loop), we choose a random edge xy, remove uv, xy, and with probability $1/2$ add ux, vy; otherwise, add uy, vx.

We already saw such switching in Section 4.4. In this application, it generates a random graph that is very close to the uniform distribution over the family of simple graphs. It should solve all problems in $O(1)$ time, provided that the graph is sparse.

Evaluating the Quality of Clustering Algorithms

Suppose that we run an algorithm on a network with known **ground truth** (such as the Zachary's karate club) or a synthetic benchmark graph (such as **LFT** or **ABCD**). The partition returned by the algorithm is most likely different than the **ground truth** partition. How then one can evaluate the quality of the algorithm? One possibility is to use a measure based on information theory, such as the **Adjusted Mutual Information** (**AMI**), that allows us to quantify the similarity between two partitions of the same set of nodes. Alternatively, one can measure similarity by considering pairs of nodes; one example of such measure is the **Adjusted Rand Index** (**ARI**). These measures are designed for comparing set partitions and not graph partitions specifically. We call them *graph-agnostic* as they ignore the graph structure. There are more sophisticated *graph-aware* measures for graph partition similarity but we do not discuss them here.

Suppose that $\mathcal{U}$ and $\mathcal{W}$ are two partitions of the same set of nodes V; in particular, $U_1 \cup U_2 \cup \ldots \cup U_u = V$ and $W_1 \cup W_2 \cup \ldots \cup W_w = V$. The mutual information of communities overlap between the two partitions can be summarized in the form of **contingency table** that is the $u \times w$ matrix $N = (n_{ij})_{i \in [u], j \in [w]}$, where n_{ij} is the number of nodes that belong to both community U_i and W_j, that is, $n_{ij} = |U_i \cap W_j|$. For $i \in [u]$ and $j \in [w]$, let $P_U(i)$, $P_W(j)$ and, respectively, $P_{UW}(i,j)$ be the probability that a random node belongs to community U_i, to community W_j and, respectively, to both communities; that is,

$$P_U(i) = \frac{|U_i|}{n}, \; P_W(j) = \frac{|W_j|}{n}, \quad \text{and} \quad P_{UW}(i,j) = \frac{|U_i \cap W_j|}{n} = \frac{n_{ij}}{n}.$$

The **Mutual Information (MI)** measures the information shared by the two community assignments:

$$MI(\mathcal{U}, \mathcal{W}) = \sum_{i \in [u], j \in [w]: P_{UW}(i,j) > 0} P_{UW}(i,j) \log_2 \left(\frac{P_{UW}(i,j)}{P_U(i) P_W(j)} \right).$$

$MI(\mathcal{U}, \mathcal{W})$ measures the information that $\mathcal{U}$ and $\mathcal{W}$ share; it tells us how much knowing one of these partitions reduces our uncertainty about the other one.

Note that $MI(\mathcal{U}, \mathcal{W}) = 0$ if $\mathcal{U}$ and $\mathcal{W}$ are independent of each other (that is, $P_{UW}(i,j) = P_U(i)P_W(j)$ for all i, j). At the other extreme, when the two partitions are identical, $MI(\mathcal{U}, \mathcal{W}) = H(\mathcal{U}) = H(\mathcal{W})$, where

$$H(\mathcal{U}) = -\sum_{i \in [u]} P_U(i) \log_2 \left(P_U(i) \right)$$

is the **Shannon entropy** associated with partition $\mathcal{U}$. In any case, $MI(\mathcal{U}, \mathcal{W})$ is upper bounded by the entropies $H(\mathcal{U})$ and $H(\mathcal{W})$. This observation justifies the following definition of the **Normalized Mutual Information (NMI)**.

Let $\mathcal{U}$ $(U_1 \cup U_2 \cup \ldots \cup U_u)$ and $\mathcal{W}$ $(W_1 \cup W_2 \cup \ldots \cup W_w)$ be two partitions of the same set of nodes V. For $i \in [u]$ and $j \in [w]$, let $n_i = |U_i|$, $n_j' = |W_j|$, and $n_{ij} = |U_i \cap W_j|$. The **Normalized Mutual Information** between $\mathcal{U}$ and $\mathcal{W}$ is defined as follows:

$$
\begin{aligned}
NMI(\mathcal{U}, \mathcal{W}) &= \frac{MI(\mathcal{U}, \mathcal{W})}{(H(\mathcal{U}) + H(\mathcal{W}))/2} \\[2mm]
&= \frac{-2 \sum_{i \in [u]} \sum_{j \in [w]} n_{ij} \log_2 \left(\frac{n \cdot n_{ij}}{n_i \cdot n_j'} \right)}{\sum_{i \in [u]} n_i \log_2(n_i/n) + \sum_{j \in [w]} n_j' \log_2(n_j'/n)}.
\end{aligned}
$$

Alternatively, one could normalize $MI(\mathcal{U}, \mathcal{W})$ by dividing it by $\min\{H(\mathcal{U}), H(\mathcal{W})\}$ or $\max\{H(\mathcal{U}), H(\mathcal{W})\}$. All of these variants have a nice feature that they range between zero when the algorithm completely fails and one when the algorithm works perfectly. Moreover, they maintain the symmetry of the mutual information with respect to its two arguments. The most widely accepted normalization is the one proposed above.

Let us now turn into pair counting based measures that built upon counting pairs of items on which two clusterings agree or disagree. As before, consider $\mathcal{U}$ and $\mathcal{W}$, two partitions of the same set of nodes V. Let Q_U and Q_W denote the set of pairs of nodes lying in one of the parts of $\mathcal{U}$ and, respectively, $\mathcal{W}$. Moreover, for a given set $Q \subseteq \binom{V}{2}$ consisting of pairs of nodes, we will use $\overline{Q} = \binom{V}{2} \setminus Q$ to denote the **complement** of Q.

The **Rand Index (RI)** is defined as the probability that $\mathcal{U}$ and $\mathcal{W}$ agree on a pair of nodes selected uniformly at random from the set of all pairs, that is,

$$RI(\mathcal{U}, \mathcal{W}) = \frac{|Q_U \cap Q_W| + |\overline{Q_U} \cap \overline{Q_W}|}{\binom{n}{2}}.$$

Alternatively, one may view the **Rand Index** as a measure of the percentage of correct decisions made by the clustering algorithm, that is,

$$RI(\mathcal{U}, \mathcal{W}) = \frac{N_{11} + N_{00}}{N_{00} + N_{10} + N_{01} + N_{11}},$$

where N_{11} is the number of true positives (pairs of nodes that are in the same community in both $\mathcal{W}$ and the reference **ground truth** partition $\mathcal{U}$), N_{10} is the number of false negatives, N_{01} is the number of false positives, and N_{00} is the number of true negatives; that is,

$$
\begin{aligned}
N_{11} &= \sum_{i \in [u]} \sum_{j \in [w]} \binom{n_{ij}}{2} \qquad\qquad\qquad (5.4)\\
N_{10} &= \sum_{i \in [u]} \binom{n_i}{2} - N_{11} \\
N_{01} &= \sum_{j \in [w]} \binom{n'_j}{2} - N_{11} \\
N_{00} &= \binom{n}{2} - N_{11} - N_{10} - N_{01} \\
&= \binom{n}{2} + N_{11} - \sum_{i \in [u]} \binom{n_i}{2} - \sum_{j \in [w]} \binom{n'_j}{2}.
\end{aligned}
$$

Combining all of these things together, we get our next definition.

Let $\mathcal{U}$ $(U_1 \cup U_2 \cup \ldots \cup U_u)$ and $\mathcal{W}$ $(W_1 \cup W_2 \cup \ldots \cup W_w)$ be two partitions of the same set of nodes V. For $i \in [u]$ and $j \in [w]$, let $n_i = |U_i|$, $n'_j = |W_j|$, and $n_{ij} = |U_i \cap W_j|$. The **Rand Index** between $\mathcal{U}$ and $\mathcal{W}$ is defined as follows:

$$RI(\mathcal{U}, \mathcal{W}) = \frac{2\sum_{i\in[u]}\sum_{j\in[w]}\binom{n_{ij}}{2} - \sum_{i\in[u]}\binom{n_i}{2} - \sum_{j\in[w]}\binom{n'_j}{2} + \binom{n}{2}}{\binom{n}{2}}.$$

The above mentioned similarity measures fail to satisfy the following, rather intuitive and desired, property known as the **constant baseline property**. In order to understand it, suppose that $\mathcal{U}$ is the ground truth partition of V. For two given natural numbers i, j such that $i > j \geq 2$, let $\mathcal{W}_i$ and $\mathcal{W}_j$ be two partitions sampled independently and uniformly at random from the set of partitions into i and, respectively, j parts. One would expect that the similarity measure between $\mathcal{U}$ and $\mathcal{W}_i$ and between $\mathcal{U}$ and $\mathcal{W}_j$ should be the same, ideally equal to zero. However, this is not the case for the **Mutual Information** (including the normalized counterpart) nor the **Rand Index** that tends to increase together with i. This is the main reason why these measures are mainly used in its adjusted form.

In order to correct the measures for randomness, it is necessary to specify a model according to which random partitions are generated. One commonly used model is the **generalized hypergeometric model** in which partitions are generated randomly, subject to having a fixed number of parts and the number of elements in each part.

In this model, the expected mutual information between two random partitions $\mathcal{U}$ and $\mathcal{W}$ is equal to

$$\mathbb{E}[MI(\mathcal{U}, \mathcal{W})] = \sum_{i\in[u]}\sum_{j\in[w]}\sum_{n_{ij}=\max\{n_i+n'_j-n,1\}}^{\min\{n_i,n'_j\}} \frac{n_{ij}}{n}\log_2\left(\frac{n\cdot n_{ij}}{n_i\cdot n'_j}\right)$$
$$\times \frac{n_i!\,n'_j!\,(n-n_i)!\,(n-n'_j)!}{n!\,n_{ij}!\,(n_i-n_{ij})!\,(n'_j-n_{ij})!\,(n-n_i-n'_j+n_{ij})!}.$$

The **Adjusted Mutual Information (AMI)** is then defined as follows:

$$AMI(\mathcal{U}, \mathcal{W}) = \frac{MI(\mathcal{U}, \mathcal{W}) - \mathbb{E}[MI(\mathcal{U}, \mathcal{W})]}{\max\{H(\mathcal{U}), H(\mathcal{W})\} - \mathbb{E}[MI(\mathcal{U}, \mathcal{W})]}.$$

As a result, the **AMI** is equal to one when the two partitions are identical and equal to zero when the **MI** between two partitions equals the value expected due to chance alone.

Exactly the same strategy can be applied to adjust the **Rand Index**. Under the generalized hypergeometric model, it can be shown that

$$\mathbb{E}\left[\sum_{i\in[u]}\sum_{j\in[w]}\binom{n_{ij}}{2}\right] = \frac{\sum_{i\in[u]}\binom{n_i}{2}\sum_{j\in[w]}\binom{n'_j}{2}}{\binom{n}{2}}.$$

Hence, the **Adjusted Rand Index (ARI)** can be conveniently simplified as follows:

$$ARI(\mathcal{U}, \mathcal{W}) = \frac{\sum_{i \in [u]} \sum_{j \in [w]} \binom{n_{ij}}{2} - \left(\sum_{i \in [u]} \binom{n_i}{2} \sum_{j \in [w]} \binom{n'_j}{2} \right) / \binom{n}{2}}{\left(\sum_{i \in [u]} \binom{n_i}{2} + \sum_{j \in [w]} \binom{n'_j}{2} \right) / 2 - \left(\sum_{i \in [u]} \binom{n_i}{2} \sum_{j \in [w]} \binom{n'_j}{2} \right) / \binom{n}{2}}.$$

Alternatively, using the notation introduced in (5.4),

$$ARI(\mathcal{U}, \mathcal{W}) = \frac{2(N_{00} N_{11} - N_{01} N_{10})}{(N_{00} + N_{01})(N_{01} + N_{11}) + (N_{00} + N_{10}).(N_{10} + N_{11})}.$$

5.4 Graph Modularity

The key ingredient for many clustering algorithms is **modularity** that we discuss in this section. As already mentioned earlier, modularity is at the same time a global criterion to define communities, a quality function of community detection algorithms, and a way to measure the presence of community structure in a network. Modularity for graphs is based on the comparison between the actual density of edges inside a community and the density one would expect to have if the nodes of the graph were attached at random, regardless of community structure, while respecting the nodes' degree on average. Such reference random family of graphs (known in this context as the **null-model**) is the **Chung-Lu** random model that we introduced in Section 2.5. Despite some known issues with the modularity function, such as the "resolution limit" that we discuss later on, many popular algorithms for partitioning large graph data sets use it.

Definition

Let $G = (V, E)$ be a graph on the set of nodes $V = \{v_1, v_2, \ldots, v_n\}$. Let

$$\mathbf{d} = \left(\deg(v_1), \deg(v_2), \ldots, \deg(v_n) \right)$$

be the degree sequence of G. For a given partition $\mathcal{A} = \{A_1, A_2, \ldots, A_\ell\}$ of V, the **modularity function** is defined as follows:

$$
\begin{aligned}
q_G(\mathcal{A}) &= \frac{1}{|E|} \sum_{A_i \in \mathcal{A}} \left(e_G(A_i) - \mathbb{E}_{G' \sim \mathcal{G}(\mathbf{d})} [e_{G'}(A_i)] \right) \\
&= \sum_{A_i \in \mathcal{A}} \frac{e_G(A_i)}{|E|} - \sum_{A_i \in \mathcal{A}} \frac{\mathbb{E}_{G' \sim \mathcal{G}(\mathbf{d})} [e_{G'}(A_i)]}{|E|},
\end{aligned}
\tag{5.5}
$$

where $e_G(A_i) = |\{v_j v_k \in E : v_j, v_k \in A_i\}|$ is the number of edges in the subgraph of G induced by set A_i. The modularity measures the deviation of the number of edges of G that lie inside parts of $\mathcal{A}$ from the corresponding expected value based on the **Chung-Lu** random graph $\mathcal{G}(\mathbf{d})$, the random graph with expected degree sequence $\mathbf{d}$. The expected value for part A_i can be computed as follows:

$$
\mathbb{E}_{G' \sim \mathcal{G}(\mathbf{d})}[e_{G'}(A_i)] \;=\; \sum_{v_j v_k \in \binom{A_i}{2}} \frac{\deg(v_j)\deg(v_k)}{2|E|} + \sum_{v_j \in A_i} \frac{\deg^2(v_j)}{4|E|}
$$

$$
= \; \frac{1}{4|E|} \left(\sum_{v_j \in A_i} \deg(v_j) \right)^2 = \frac{(\mathrm{vol}(A_i))^2}{4|E|}.
$$

The first term in (5.5), $\sum_{A_i \in \mathcal{A}} e_G(A_i)/|E|$, is called the **edge contribution**, whereas the second one, $\sum_{A_i \in \mathcal{A}} (\mathrm{vol}(A_i))^2/4|E|^2$, is called the **degree tax**. It is easy to see that both terms are between zero and one and so, in particular, $q_G(\mathcal{A}) \leq 1$. Also, if $\mathcal{A} = \{V\}$, then $q_G(\mathcal{A}) = 0$, and if $\mathcal{A} = \{\{v_1\}, \ldots, \{v_n\}\}$, then $q_G(\mathcal{A}) = -\frac{\sum \deg^2(v)}{4|E|^2} < 0$. On the other hand, it can be shown that $q_G(\mathcal{A}) \geq -1/2$.

The maximum **modularity** $q^*(G)$ is defined as the maximum of $q_G(\mathcal{A})$ over all possible partitions $\mathcal{A}$ of V; that is, $q^*(G) = \max_{\mathcal{A}} q_G(\mathcal{A})$. Despite the fact $q^*(G)$ is well defined, as mentioned in Section 5.2, the number of partitions of V increases very fast with $|V|$ and so, in practice, one is not able to find a partition that maximizes the modularity function unless graph G is really small. Hence, all algorithms that try to optimize the modularity function are heuristic in nature. In order to maximize $q_G(\mathcal{A})$ one wants to find a partition with large edge contribution subject to small degree tax. If $q^*(G)$ approaches 1 (which is the trivial upper bound), we observe a strong community structure; conversely, if $q^*(G)$ is close to zero (which is the trivial lower bound), there is no community structure. Finally, let us mention that the definition in (5.5) can be easily generalized to weighted graphs by replacing edge counts with sums of edge weights. Adjusting it to directed graphs is also straightforward and we do it at the end of this section.

Algorithms

Since scalability is an important issue in all community detection algorithms, almost all modularity based algorithms follow some greedy optimization methods. For example, one of the earlier attempts include the **CNM** algorithm (**Clauset–Newman–Moore**) that merges two communities whose amalgamation produces the largest increase in modularity function. Here we provide more details on another greedy algorithm, namely, the **Louvain** algorithm that is, arguably, the best algorithm from this class of algorithms. It appears to run in time $O(n \ln^2 n)$ where n is the number of nodes in the network.

In this algorithm, small communities are first found by optimizing modularity locally on all nodes. Then, each small community is grouped into one node and the original step is repeated on a smaller graph. The process stops when no improvement on the modularity function can be further achieved.

One pass of the **Louvain** algorithm consists of two phases that are repeated iteratively. Initially, each node in the network is assigned to its own community. For each node v, we consider all neighbours u of v and compute the change in the modularity function if v is removed from its own community and moved into the community of u. It is important to mention that this value can be easily and efficiently calculated without the need to recompute the modularity function from scratch. Once all the communities that v could belong to are considered, v is placed into the community that resulted in the largest increase of the modularity function. If no increase is possible, v remains in its original community. The process is repeated for the remaining nodes following a given (typically random) permutation of nodes. If no increase is possible after considering all nodes, a local maximum value is achieved and the first phase ends.

During the second phase, the algorithm contracts all nodes that belong to one community into a single node. All edges within that community are replaced by a single weighted loop. Similarly, all edges between two communities are replaced by a single weighted edge. Once the new network is created, the second phase ends. The resulting graph is typically much smaller than the original graph. As a result, the first pass is typically the most time consuming part of the algorithm.

Despite the fact that the **Louvain** algorithm offers one of the best trade-offs between the quality of the clusters it produces and its speed, it has some stability issues. This is mainly seen in unweighted graphs, where the randomized order of the nodes in a pass can lead to very different communities; this is due to the fact that there can be several community moves that lead to the same change in the modularity function.

In order to produce more stable outcomes, one may use the concept of ensemble clustering often referred to as consensus clustering. The main idea behind this concept is to combine several partitions over the same dataset to produce a final, more stable partition.

Let $G = (V, E)$ be any weighted graph on n nodes. The **Ensemble Clustering algorithm for Graphs (ECG)** starts with ℓ randomized level-1 partitions $\mathcal{P}_1, \mathcal{P}_2, \ldots, \mathcal{P}_\ell$ of the set of nodes that are obtained by ℓ independent runs of the first phase of the **Louvain** algorithm and only once. (We do not discuss complexity issues in length but let us point out that this step can be easily distributed as these ℓ runs are completely independent.) Running a single level of the algorithm is a good example of a weak learner, where nodes are grouped into many small clusters. We assign new weights to edges of G as follows: for each $uv \in E$,

$$
w(uv) = \begin{cases} w_* + (1 - w_*) \cdot \frac{1}{\ell} \sum_{i=1}^{\ell} \alpha_i(u, v), & \text{if } uv \text{ belongs to the 2-core of } G, \\ w_* & \text{otherwise,} \end{cases}
$$

where $w_* \in [0,1]$ is the parameter of the algorithm and $\alpha_i(u,v) = 1$ if u and v appear together in some part of partition $\mathcal{P}_i$; otherwise, $\alpha_i(u,v) = 0$. Clearly, $w(uv) \in [w_*, 1]$ and the minimum weight w_* is assigned to edges that were never put into the same part in any partition $\mathcal{P}_i$, or are outside of the 2-core. Finally, the **Louvain** algorithm is performed (till the very end, *not* just level-1) on a weighted version of the initial graph $G = (V, E)$. Note that the original weights of the edges of G are discarded but, of course, they were taken into account when creating partitions $\mathcal{P}_i$.

When running the **ECG** algorithm, the size ℓ of the ensemble and the minimum edge weight w_* are the only parameters that need to be supplied. However, the results are not too sensitive to their choice (assuming ℓ is large enough and w_* is relatively close to 0) and the default values are usually suitable, namely, $\ell = 16$ and $w_* = 0.05$.

Another issue with the **Louvain** algorithm is that it may produce badly connected communities, in the extreme situations the communities it produces could even be disconnected. The **Leiden** is a recently proposed modification of the **Louvain** algorithm that, among other things, ensures that communities are connected. This algorithm is implemented in `igraph`, and is also available as an option with **ECG**.

Resolution Limit

Recall that the modularity function compares the number of edges within one part of a partition with the expected number of edges one would see in the network with the same degree distribution but edges wired randomly. Such random null-model implicitly assumes that each node in a network can be potentially adjacent to any other node. This assumption is not reasonable in practice, especially if the network at hand is large. Unfortunately, it has an important implication known in the literature as the **resolution limit** that can be summarized as follows: optimizing modularity function in large networks cannot find small communities, even if they are well defined.

Indeed, the expected number of edges between part A_i and part $A_j \neq A_i$ in the corresponding **Chung-Lu model** $\mathcal{G}(\mathbf{d})$ is equal to

$$\sum_{v_i \in A_i} \sum_{v_j \in A_j} \frac{\deg(v_i)\deg(v_j)}{2|E|} = \frac{\mathrm{vol}(A_i)\,\mathrm{vol}(A_j)}{2|E|}.$$

Hence, if $\mathrm{vol}(A_i) \leq \mathrm{vol}(A_j) < \sqrt{2|E|}$, then the expected number of edges between the two parts is smaller than one. If this happens, then a single edge between them would be interpreted by algorithms that use the modularity function that there is a correlation between them and would lead to merging the corresponding two parts. In particular, even weakly interconnected complete graphs that have the largest possible internal density and represent clearly identifiable communities, would be merged together, provided the network is sufficiently large.

In order to deal with the resolution limit, one may use the original **Louvain** algorithm to find an initial partition $\mathcal{P} = (A_1, A_2, \ldots, A_\ell)$ of the entire network. The goal is then to try to subdivide some large communities A_i by independently running the algorithm on $G[A_i]$, the graph induced by part A_i. Alternatively, one may run **ECG** that tends to deal better with this issue. In particular, for the ring of cliques mentioned above, it gives the weight $w(uv)$ close to one for edges within cliques but close to w_* for edges between them, thus reducing the risk of merging cliques. In Section 5.7, we experiment with the ring of cliques and discuss this issue more.

Yet another solution, known as **multiresolution method**, adds a resistance $r \in \mathbb{R}$ to every node that can be viewed as a loop of weight r. Positive value of r increases the aversion of nodes to form communities whereas negative value of r does the opposite. Independently, one may multiply the degree tax by a universal constant $\gamma \in \mathbb{R}_+$.

Directed Graphs

Generalizing the modularity function to directed graphs (including weighted directed graphs) is straightforward. As before, the goal is to compare the number of directed edges of D that lie inside some part of a given partition with the corresponding expected value based on the corresponding null-model. In order to achieve it, one needs to generalize the **Chung-Lu** random graph $\mathcal{G}(\mathbf{d})$, the random graph with expected degree sequence $\mathbf{d}$, to directed graphs.

Let $D = (V, E)$ be a directed graph on the set of nodes $V = \{v_1, v_2, \ldots, v_n\}$. Let

$$
\begin{aligned}
\mathbf{d}^{in} &= \left(\deg^{in}(v_1), \deg^{in}(v_2), \ldots, \deg^{in}(v_n) \right) \\
\mathbf{d}^{out} &= \left(\deg^{out}(v_1), \deg^{out}(v_2), \ldots, \deg^{out}(v_n) \right)
\end{aligned}
$$

be the degree sequence of D. For a given partition $\mathcal{A} = \{A_1, A_2, \ldots, A_\ell\}$ of V, the **modularity function** is defined as follows:

$$
q_D(\mathcal{A}) = \sum_{A_i \in \mathcal{A}} \frac{e_D(A_i)}{|E|} - \sum_{A_i \in \mathcal{A}} \frac{(\text{vol}^{in}(A_i))(\text{vol}^{out}(A_i))}{|E|^2},
$$

where $e_D(A_i) = |\{v_j v_k \in E : v_j, v_k \in A_i\}|$ is the number of directed edges in the subgraph of D induced by set A_i, $\text{vol}^{in}(A_i) = \sum_{v \in A_i} \deg^{in}(v)$, and $\text{vol}^{out}(A_i) = \sum_{v \in A_i} \deg^{out}(v)$. As before, the first term is called the **edge contribution** whereas the second one is called the **degree tax**. All algorithms that optimize the modularity function can be now reused to deal with directed graphs.

5.5   Hierarchical Clustering

Graph clustering algorithms are *unsupervised* machine learning tools. In particular, it is usually assumed that the decision about the number of the communities the set of nodes should be partitioned into should be made by the algorithm. This is in contrast to, for example, k-**means** clustering algorithm of data points in the form of d-dimensional vectors in which the number of clusters, k, is provided as a parameter. Independently, other algorithms may be run on top of k-**means** to select the number of clusters.

Alternatively, similarly to standard hierarchical clustering algorithms developed for tabular data (that is, data that is structured into rows, each of which contains information about some feature), one may apply a general method of building a hierarchy of clusters. Strategies for such **hierarchical clustering** algorithms generally fall into two natural types: agglomerative or divisive. **Agglomerative** algorithms start with the trivial partition of the set of nodes V into n clusters, where each cluster consists of a single node. At each step of the process, an algorithm selects a pair of clusters with the largest similarity and then these two clusters are merged into one. The process continues until we reach the other extreme trivial partition of V with only one cluster, V itself. Instead of this "bottom-up" approach, one may consider a "top-down" approach that is used by the second family of hierarchical algorithms. Indeed, **divisive** algorithms start with the trivial partition consisting of a single cluster including the entire set V and then they try to refine the partition by splitting one of the clusters into two. The process ends with the partition in which each node is a singleton and forms its own cluster.

Regardless which of the two types of hierarchical clustering we use, the outcome will be the sequence of n partitions, one for each number of parts. Such outcomes can be conveniently represented by a **dendrogram** known also as a **hierarchical tree**. (See Figure 5.4 in the section with experiments for an example of a dendrogram.) On the left side we usually put the leaves of the tree that correspond to the labels of the nodes of a graph. Going right, each time exactly one pair of clusters is merged which is indicated by a vertical line joining two vertical lines corresponding to the two clusters.

Hierarchical clustering algorithms provide an interesting point of view but they have some clear limitations. First of all, they provide a sequence of n partitions and it is often not clear which one should be used for a given network and problem at hand. One natural approach is to benchmark these partitions using some external criterion (such as the modularity function we discussed in Section 5.4) and select the partition that yields the best "score." This is what the **Girvan–Newman** algorithm does by default. Moreover, these algorithms are rather slow and so they cannot be used for large networks that we often need to deal with in practice. Finally, note that the outcome of the process heavily depends on the similarity measure (for agglomerative algorithms) or dissimilarity measure (for divisive ones) between clusters. Below, we present some specific implementations, one for each type, but there are many other natural measures one may want to use. We discuss this more and provide other alternatives in Section 6.2 in which similarity between nodes becomes important again, this time to embed the nodes of a graph.

Ravasz (Agglomerative) Algorithm

In order to illustrate agglomerative algorithms, we consider the following particular implementation, **Ravasz algorithm**, that uses the topological overlap matrix to measure similarity between nodes of an unweighted graph G. The idea behind it is to design a measure that is large for pairs of nodes that belong to the same community and small otherwise.

Let $G = (V, E)$ be any unweighted graph on the set of nodes $V = \{v_1, v_2, \ldots, v_n\}$ without any isolated nodes. The **topological overlap matrix** $\mathbf{S} = (s(v_i, v_j))_{i,j \in [n]}$ is a quadratic and symmetric matrix in which

$$s(v_i, v_j) = \frac{|N(v_i) \cap N(v_j)| + \delta_{v_i v_j \in E}}{\min\{\deg(v_i), \deg(v_j)\} + \delta_{v_i v_j \notin E}}.$$

In the formula above, δ_A is the **Kronecker delta**: $\delta_A = 1$ if A holds and $\delta_A = 0$ otherwise. Let us point out that always $s(v_i, v_j) \in [0, 1]$. Moreover, $s(v_i, v_j) = 1$ if and only if nodes v_i and v_j are adjacent and all neighbours of v_i (other than v_j) are also adjacent to v_j (or vice versa). On the other extreme, $s(v_i, v_j) = 0$ if and only if the corresponding nodes have no common neighbours and are not adjacent.

More importantly, matrix $\mathbf{S}$ provides a way to measure similarity between nodes but during the algorithm, one needs to evaluate how similar two communities are. Suppose that we are given two non-overlapping communities, A and B; that is, $\emptyset \neq A \subseteq V$, $\emptyset \neq B \subseteq V$, and $A \cap B = \emptyset$. The three natural choices are: **single**, **complete** and **average cluster similarity**, $s_s(A, B)$, $s_c(A, B)$ and, respectively, $s_a(A, B)$ defined as follows:

$$\begin{aligned}
s_s(A, B) &= \min\{s(a, b) : a \in A, b \in B\} \\
s_c(A, B) &= \max\{s(a, b) : a \in A, b \in B\} \\
s_a(A, B) &= \frac{1}{|A||B|} \sum_{a \in A, b \in B} s(a, b).
\end{aligned}$$

The **Ravasz algorithm** uses the average.

As it is done in all agglomerative algorithms, this particular instance initially assigns each node to its own community and computes the topological overlap matrix by evaluating $s(v_i, v_j)$ for all $\binom{n}{2}$ pairs of nodes. Then, one needs to identify a pair of communities with the largest similarity and merge them into a single community (in case of a tie, a decision is made randomly). After that operation, the similarity matrix has to be updated (as noted above, we store the similarities between communities in a matrix) which can be done by computing similarities between the new community and all remaining communities. We repeat this process until we reach the trivial partition consisting of the whole set of nodes. Both the time and the space complexity of this algorithm are $\Theta(n^2)$, provided that the average node degree in the graph is $\Theta(1)$, which is reasonable for small graphs but, as already mentioned, not scalable.

Girvan–Newman (Divisive) Algorithm

In order to illustrate divisive algorithms, we consider the following particular implementation, **Girvan–Newman algorithm**. Later on, we will discuss another example of such algorithms using matrix algebra called **Spectral Bisection Method**. The goal of **Girvan–Newman algorithm** is to systematically remove edges between nodes that we suspect that they belong to different communities. From time to time, this operation disconnects some connected component of a graph that is split into two, resulting in another branch in the corresponding dendrogram. The goal now is to design a measure between a pair of nodes that is large for pairs of nodes that belong to different communities and small otherwise.

The idea used is closely related to **betweenness centrality** discussed in Section 3.3. Recall that node betweenness is an indicator of highly central nodes in a graph. Indeed, if many shortest paths go through node v, then v plays a central role within the network. This notion naturally extends to edges. For a given edge uv, the **edge betweenness** $\ell(uv)$ is the number of shortest paths between pairs of nodes that run along edge uv. If there is more than one shortest path between a pair of nodes, each path is independently considered. If a graph consists of communities that are only loosely connected by a few edges between them, then all shortest paths between nodes in different communities must go along one of these few edges. By averaging argument, one of such edges must have large edge betweenness. (Consider, for example, the George Washington Bridge, spanning the Hudson River between New York City and Fort Lee, New Jersey, that is the world's busiest bridge in terms of vehicular traffic.)

Now, we are ready to explain details of the algorithm. We start from the original graph G and compute the edge betweenness $\ell(uv)$ of each edge $uv \in E$. In each step of the process, we remove an edge with the largest betweenness, making a random decision in case of a tie. After that we need to recompute the edge betweenness for the new graph. These steps are repeated until we are left with an empty graph on n nodes; each node belongs to its own community. The complexity of the fastest algorithm to compute betweenness of m edges is $\Theta(mn)$. Hence the worst-case complexity of the **Girvan–Newman**

algorithm is $\Theta(m^2 n)$. However, after one edge of the graph is removed, the algorithm has to recalculate the betweenness only of those edges that were affected by the removal, which is at most those that are in the same component as the removed edge. Hence, in practice, the complexity can be expected to be better than the above worst-case scenario. In any case, as before, the conclusion is that this algorithm is not scalable and so it cannot be used for large graphs we usually need to deal with.

5.6   A Few Other Methods

In previous sections, we grouped clustering algorithms into two families: algorithms based on the modularity function (Section 5.4) and hierarchical clustering (Section 5.5). There are many other approaches worth highlighting and our exposition is by no means exhaustive. In this section we present three more algorithms that use different techniques and ideas than the ones discussed earlier.

Label Propagation Algorithm

Let $G = (V, E)$ be an undirected graph. The first approach, the **Label Propagation** algorithm, tries to make local decisions based on the assumption that there are more edges present within ground-truth communities in comparison to the number of edges between communities. The algorithm starts with a trivial partition of V into $n = |V|$ communities, that is, each node belongs to its own community. Each node will keep a label representing its community; hence, there are initially n labels. In each phase of the algorithm, we investigate nodes in an order selected uniformly at random from the set of all permutations of V. Each investigated node v adjusts (if needed) its label to the label that the majority of neighbours of v have. If there are at least two labels present in the neighbourhood of v with the same maximum number of occurrences, then v adopts one of them uniformly at random. Some labels quickly propagate throughout the network whereas some other disappear. The intuition behind the algorithm is that highly connected groups of nodes should quickly reach consensus on one of the labels and then such communities should expand, affecting other parts of the graph.

The algorithm stops if at the end of a given phase each node has a label that is consistent with the majority label of its neighbours. Note that if this is the case, then a stationary state is reached and continuing the process will not change anything. Unfortunately, it is not guaranteed that such stationary state is reached and it is possible that the algorithm indefinitely cycles through the same sequence of labels. As a result, in practice, the algorithm runs for a predefined number of phases (say, 100) unless it earlier reaches a stationary state.

The **Label Propagation** algorithm is very fast. However, on a negative note, let us mention it is quite unstable and results can vary a lot at each run. This is, of course, a common problem of clustering algorithms and one may use some variant of a consensus clustering to obtain more stable variant of this algorithm. For example, in order to find such consensus, one may generate m partitions $P_1, P_2, \ldots, P_m$ by independently running the **Label Propagation** algorithm. Then, for a given measure of clustering quality we discussed earlier (for example, the **Adjusted Mutual Information** (**AMI**)), the consensus partition P^c is identified that maximizes $\sum_{i=1}^{m} AMI(P^c, P_i)$.

Spectral Bisection Method

In this section we assume that a graph $G = (V, E)$ on n nodes is simple, undirected, and connected; if G is not connected, then one may independently find communities in each connected component. Assume that our goal is to partition the set of nodes $V = \{v_1, v_2, \ldots, v_n\}$ into two subsets. We will use $x_i \in \{-1, 1\}$ to describe such partition: if $x_i = x_j$ for some $i \neq j$, then v_i and v_j belong to the same part. Indeed, this rule yields two equivalence classes that partition the set of nodes V. Let $\mathbf{x}$ denote a column vector consisting of x_i.

The model we use employs the **graph Laplacian**, that is, matrix $\mathbf{L} = \mathbf{D} - \mathbf{A}$, where $\mathbf{A} = (a(v_i, v_j))_{i,j \in [n]}$ is the (symmetric) adjacency matrix and $\mathbf{D} = (d(v_i, v_j))_{i,j \in [n]} = \mathrm{diag}(\mathbf{A1})$ is the diagonal matrix with $d(v_i, v_i) = \deg(v_i) = \sum_{j=1}^{n} a(v_i, v_j)$. Our objective is to find a partition (vector $\mathbf{x}$) that minimizes the number of edges between the two clusters. Using the fact that $(x_i - x_j)^2 = 4$ if the two corresponding nodes belong to two different clusters and otherwise $(x_i - x_j)^2 = 0$, this number can be conveniently written as:

$$L(\mathbf{x}) = \sum_{i=1}^{n-1} \sum_{j=i+1}^{n} a(v(i), v(j))(x_i - x_j)^2/4 = \mathbf{x}^T \mathbf{L} \mathbf{x}/8.$$

Let us note, however, that the number of edges between the clusters (and so equivalently the formula above) is trivially minimized when all nodes are assigned to a single cluster (that is, all x_i's are equal to 1 or all are equal to -1). Therefore, we insist that both clusters have sizes which are as close to being equal as possible, which can be expressed as $|\sum_{i=1}^{n} x_i| \leq 1$ (that is, the sum is equal to 1 or -1 if n is odd, and equal to 0 if n is even).

Unfortunately, minimizing $L(\mathbf{x})$ subject to $|\sum_{i=1}^{n} x_i| \leq 1$ when x_i are restricted to be in $\{-1, 1\}$ is a difficult combinatorial problem. Therefore, we transform it to a simpler problem that can be solved more easily and then *round* its solution to get the desired values of x_i. The relaxation that we do is to allow x_i to be any real number. This auxiliary problem can be written as finding the minimum of $L(\mathbf{x})$ subject to $\mathbf{1}^T\mathbf{x} = 0$ and $\mathbf{x}^T\mathbf{x} = n$ in real numbers (note that when $x_i \in \{1, -1\}$ we have that $\mathbf{x}^T\mathbf{x} = n$). The rounding procedure is simple: each solution of the auxiliary problem is rounded to 1 or -1 whichever is closer; for completeness, if $x_i = 0$, then 1 or -1 is assigned randomly (note that in practice vector $\mathbf{x}$ is computed numerically and it is extremely unlikely that $x_i = 0$ for some i). Let us note that this operation is equivalent to taking a sign of the solution obtained from the relaxed optimization problem.

Note that the solution that we have found satisfies $x_i \in \{-1, 1\}$ but does not have to satisfy $|\sum_{i=1}^{n} x_i| \leq 1$. Moreover, in general, it does not have to minimize $L(\mathbf{x})$ subject to this condition. However, in practice, it is a reasonable and good approximation. Alternatively, one may take this solution as a starting point and pass it to some discrete optimization algorithm to improve it but, in practice, this is rarely done.

Let us come back to the relaxed optimization problem. In order to solve it, let us note that $\mathbf{L}$ is symmetric and semi-positive definite. As a result, it has non-negative eigenvalues and its eigenvectors are orthogonal. Observe that this matrix (since it is assumed that the graph is connected) has exactly one eigenvalue equal to 0 with the associated eigenvector $\mathbf{1}$. This eigenvector is associated with the trivial partition that we ruled out, that is, in which all nodes fall into one part. However, observe that all other eigenvectors of $\mathbf{L}$ are orthogonal to $\mathbf{1}$ so they satisfy the condition $\mathbf{1}^T x = 0$ and have a strictly positive norm. Therefore, using the linear algebra theory on eigendecomposition of the matrix, we may take a properly scaled eigenvector associated with the second smallest eigenvalue of $\mathbf{L}$ (which we know is strictly positive) and it will be a minimizer of $L(\mathbf{x})$ subject to $\mathbf{1}^T\mathbf{x} = 0$ and $\mathbf{x}^T\mathbf{x} = n$, as needed. The vector $\mathbf{x}$ we have just found is often called the **Fiedler vector** and the second smallest eigenvalue of $\mathbf{L}$ is called the **Fiedler value**.

In summary, the **Spectral Bisection Method** is conducted in three steps. First, one needs to find matrix $\mathbf{L}$; note that in practice we do not have to fully materialize it as it is most likely very sparse. Next, one needs to find the eigenvector corresponding to the second smallest eigenvalue. Finally, signs of the entries of this eigenvector yield the assignment of nodes to the two communities. Note that this procedure can be then applied recursively to find a more fine grained clustering.

Let us point out a few properties of the **Spectral Bisection Method**. Note that it is an example of a wider class of community detection procedures in which one first finds an embedding of nodes of a graph (we discuss embeddings in Chapter 6) and then clustering is performed using one of the traditional machine learning algorithms. As a practical comment, let us note that the second smallest eigenvalue does not have to have a unique eigenvector associated with it, that is, its multiplicity might be greater than 1. In such case, any linear combination of these vectors is a minimizer of the relaxed problem; fortunately, since graphs are large and have irregularities in their structure,

such undesired situation is highly unlikely in practice. Let us also note that
the optimization we performed is quite similar to modularity except that we
restricted ourselves to two parts and we insisted that they are well balanced.
If one wants to find parts with similar volumes instead of similar number of
nodes, then in the optimization procedure one should insist that $\mathbf{1}^T\mathbf{Dx} = 0$
(instead of $\mathbf{1}^T\mathbf{x} = 0$). Such approach would use one-dimensional **Laplacian
Eigenmaps** embedding (**LEM**) as a solution to the relaxed problem and we
describe it in Section 6.3.

Finally, in order to see the reason one might not want to ignore the degrees
of nodes in the partitioning, let us consider the following example. Take sufficiently large even value of n and consider two complete graphs on n nodes;
graph G has nodes labelled $g_1, g_2, \ldots, g_n$ and graph H has nodes labelled
$h_1, h_2, \ldots, h_n$. Take their union and additionally add edges $g_i h_i$ and $g_i h_{n-i}$
($i \in [n]$). Finally, add n nodes g_i' that are connected only to g_i and h_i' that
are connected only to h_i ($i \in [n]$). There are two partitions that minimize the
number of edges between parts. The first one puts all g_i and h_i nodes in one
cluster and all g_i' and h_i' into the second cluster. The second partition puts
all g_i and g_i' nodes in one cluster and all h_i and h_i' into the second cluster.
Intuitively, the first partition is very bad while the second partition is good.
If the volumes are considered instead of part sizes, the latter cluster would be
preferred. Fortunately, in this particular case, **Spectral Bisection Method**
as well as one-dimensional **LEM** recover the desired clustering as preferred
one (so, actually, in this case the relaxation helps since embedding nodes on
$\mathbb{R}$ captures the similarity of nodes better than embedding on $\{-1, 1\}$).

Infomap

The next community detection algorithm uses yet another approach to detect
communities. It uses concepts taken from the information theory and hence its
name **Infomap**. Having said that, the optimization framework is very similar
to what is done in the **Louvain** algorithm discussed in Section 5.4. As in the
original **Louvain** algorithm, one starts with a trivial partition of the set of
nodes and tries to optimize the well defined quality function (in this case, the
entropy function that we will define soon) using similar heuristic optimization
strategies.

Let us consider a relatively long **random walk** performed on a given undirected graph $G = (V, E)$. The sequence of nodes that are visited by the walk
preserves important information about the structure and the topology of the
network. In particular, since there are more edges within ground-truth communities and relatively fewer edges between them, a random walk on a graph
with strong community structure often gets trapped within one community,
staying there for a long time before moving to the next community where it
gets stuck for a while again.

Using this intuition, one may benchmark a given partition $\mathcal{A} =
\{A_1, A_2, \ldots, A_\ell\}$ of the set of nodes V as follows. The goal is to uniquely

represent the sequence of nodes visited by the random walk by a sequence of zeros and ones. In this process, we need to label nodes in G but we are also allowed to use two additional labels assigned to each community A_i: the *entry label* indicating that the walk entered A_i and the *exit label* indicating that the walk left A_i. Each label (assigned to nodes, entry, and exit labels) is a bit sequence and labels do not have to have the same length. In particular, if some node within a community is visited very often by a random walk, then it should get a shorter label than some other node in this community that is visited less often. The benefit of using entry and exit labels is to allow the same label to be assigned to many nodes as long as they belong to different communities. As a result, the encoding of the walk should be shorter, provided that there is some strong community structure in the network, as one has to "pay" the cost of encoding entry and exit labels. The exit label needs to be distinct from the node labels within the same community but it can be the same as label of some other nodes or exit/entry labels, as long as there is no ambiguity in reconstructing the random walk.

In order to build an intuition how codes that vary lengths of labels work, let us pause our discussion on clustering for a moment and consider the **UTF-8** encoding of characters[1]. In this encoding, each character is encoded using 1, 2, 3, or 4 bytes. Regular **ASCII** characters use only 1 byte, and have the first bit always set to 0. This means that we can encode 128 characters this way following the pattern *0xxxxxxx*, where x is either 0 or 1. Characters that are encoded on two bytes follow the pattern *110xxxxx 10xxxxxx*. Note that the first bit of the sequence is 1 and so there is no risk that a two byte encoded character will be confused with one byte encoded character. Similarly, three and four byte characters are encoded as *1110xxxx 10xxxxxx 10xxxxxx* and, respectively, as *11110xxx 10xxxxxx 10xxxxxx 10xxxxxx*.

One might wonder why non leading bytes of the encoding use *10* prefix. Indeed, more compression could have been achieved if this restriction were not imposed. However, the benefit of this approach is that **UTF-8** encoded string has a property of being more resistant to errors in communication. If we see a byte starting with *10*, then we know that it is not a start of a character so we are sure we will not start decoding an **UTF-8** string from an incorrect position.

Observe also that the encoding rules make sure that when we see the start of the character, we immediately can identify the length of the encoding. It is then natural to use shorter encodings for characters that are encountered more often in the texts and longer encodings to characters that are rare. As a result, typical texts that mostly use **ASCII** characters and only sporadically non-ASCII characters are represented compactly.

Let us now come back to **Infomap**. The main assumption behind this algorithm is that good partitions (that is, those that are close to the ground-truth)

---

[1] see en.wikipedia.org/wiki/UTF-8

have the property that there *exists* a short representation of the random walk from the family of possible representations we have described above. From theoretical point of view, the length of a shortest representation is well defined but finding it might be computationally challenging. Fortunately, one may use results from information theory, in particular, the fundamental and classic result known as the **Shannon's source coding theorem** that establishes the limits to possible data compression. This result implies that a shortest possible bit-string representing the random walk has the average number of bits per one step of the walk equal to the corresponding **entropy** function that in this case is given by the following **map equation**:

$$L = L(\mathcal{A}) = q \cdot H(Q) + \sum_{i=1}^{\ell} p_i \cdot H(P_i). \tag{5.6}$$

In equation (5.6), q is the fraction of time steps the walk spends moving between parts, and p_i is the fraction of time steps it spends within part A_i and leaving it. The quantities $H(Q)$ and $H(P_i)$ are information-theoretic entropies. The **entropy** of a sequence R of r objects is given by

$$H(R) = -\sum_{i=1}^{r} R_i \log_2(R_i),$$

where R_i is the fraction of times that object i appears in the sequence. (Alternatively, one may define the entropy for the corresponding discrete random variable with r possible outcomes.) In equation (5.6), $H(Q)$ is the entropy of the sequence of entry labels and $H(P_i)$ is the entropy of the sequence of node labels and the exit label of part A_i. Hence, indeed, one may benchmark a given partition $\mathcal{A}$ without actually assigning any labels to nodes. Similarly, we do not even need to generate a random walk as one may compute q and p_i's simply by analyzing the structure of graph G. Finally, let us mention that one can perform all the required steps quickly and so the running time of **Infomap** is in practice comparable to the one of **Louvain**.

5.7 Experiments

Let us start with the Zachary karate club graph that we already experimented within this chapter. The ground-truth of this small graph consists of two communities which we illustrate in Figure 5.2. Many graph clustering algorithms, including the ones based on modularity function, often find a large number of communities as the optimal split and so are typically far away from the truth.

Given the two ground-truth communities, let us first verify if those communities satisfy the definitions of weak and strong communities introduced in

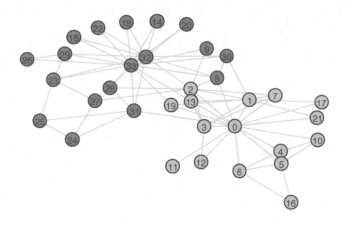

FIGURE 5.2
Zachary karate club graph with the ground truth communities.

Section 5.2. In order to check whether the two communities form strong communities, we need to compare the internal degree and external one for each node. In doing so, we find that every node satisfies the strong community condition except two nodes: node 2 has both internal and external degrees equal to 5 whereas node 9 has exactly one neighbour in each community. On the other hand, if we consider the total degree for each community, we find that one community has the total internal degree 66 and the external degree 10 while the other community has the total internal degree 70 and the external degree 10. Hence, both clearly qualify as weak communities.

We can also characterize the role of each node in the Zachary graph as described in Section 5.2. We computed the **normalized within-module degree** $z(v)$ for every node v, as well as the **participation coefficient** $p(v)$, both based on the two ground-truth communities. Given the terminology introduced in Section 5.2, there are three types of nodes in this graph: (i) **provincial hubs** ($z(v) > 2.5, p(v) < 0.3$), (ii) **peripheral non-hubs** ($z(v) \leq 2.5, 0.05 \leq p(v) < 0.62$) and (iii) **ultra-peripheral non-hubs** ($z(v) \leq 2.5, p(v) < 0.05$). In Figure 5.3(a), the provincial hub nodes are displayed as squares, the peripheral nodes as dark circles, and the ultra-peripheral nodes as pale circles. Two nodes are of special interest: node 33 is the club's president and node 0 is the instructor; recall that the two ground-truth communities are due to the split of the club with members following one of those two individuals. In Figure 5.3(b), we plot the nodes with respect to the $z(v)$ and $p(v)$ scores. Nodes 0 and 33 appear as strong hubs, while node 32 is close to the boundary between hubs and non-hubs.

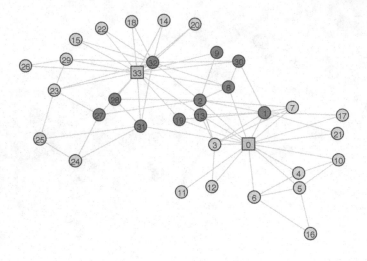

(a) graph view

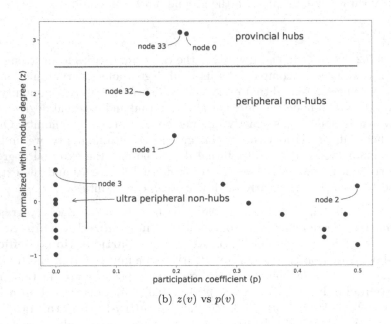

(b) $z(v)$ vs $p(v)$

FIGURE 5.3
Two different views of the node roles as characterized by the $z(v)$ and $p(v)$ scores.

Hierarchical Clustering

Next, we consider the **Girvan–Newman** hierarchical clustering algorithm. Recall that this algorithm produces a hierarchy of node partitions, and the best one is chosen according to some external criterion, such as the modularity function, of a pre-selected number of parts. This is often summarized using a visual representation called dendrogram which we show in Figure 5.4 (the node indices are represented on the left). Regardless whether we use a divisive algorithm (such as **Girvan–Newman**) or an agglomerative one (such as **Ravasz**), the dendrogram can be always read in two ways. Going from left to right in Figure 5.4, we start from a partition in which each node belongs to its own community and we gradually merge communities; for example, nodes 32 and 33 are merged first, then node 29 is added to this cluster, etc. We may also look at the dendrogram from right to left. In this case, we start with a single community consisting of all nodes and break this community up into two parts each time we reach a fork. For example, the first split divides the nodes in two parts (top and bottom group of nodes), and the next split isolates node 9 that forms its own community.

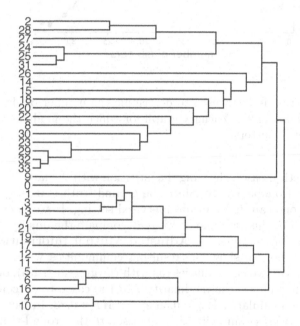

FIGURE 5.4
Dendrogram obtained with the **Girvan–Newman** hierarchical clustering algorithm performed on the Zachary graph.

Each hierarchical clustering algorithm returns n partitions, and it might be the case that the whole dendrogram is of interest. However, if one asks for a single partition, then by default the **Girvan–Newman** algorithm returns the partition from the dendrogram that yields the largest modularity function. As it is illustrated in Figure 5.5, in the case of the Zachary graph this optimum is achieved with a partition into 5 communities. Alternatively, one may simply choose the partition that yields some preselected number of clusters.

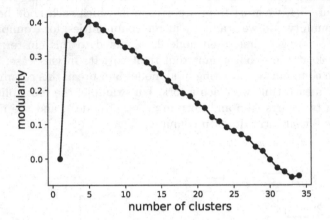

FIGURE 5.5
Modularity function is often used to determine the number of clusters. Here we show the results for the Zachary graph for which the modularity function is maximized for 5 clusters.

In many practical cases, choosing the partition with the largest modularity is a good and reasonable choice. Since the ground-truth is usually unknown, this is the best one can do in an unsupervised setting. However, we do have the ground-truth in this case and so we may compare the quality of different partitions using, for example, the **Adjusted Mutual Information (AMI)** score. In Figure 5.6, we show the resulting communities when we cut the dendrogram to have two communities (as with the ground-truth), or five communities (that maximizes the modularity). In the case of two communities, shown in (a), the modularity is $q = 0.36$ and $AMI = 0.83$. In fact, only node 2 is placed in a wrong community in comparison to the ground-truth. With 5 communities, we get $q = 0.40$ but $AMI = 0.55$. In Figure 5.6(b), we colour the nodes with respect to the ground truth, and the label on each node indicates its community when we consider highest modularity.

We repeat the above experiment on a small **ABCD** graph with 100 nodes. This graph has three ground-truth communities and so we can compare the **AMI** score and the corresponding modularity for each possible cut of the dendrogram. In Figure 5.7, we see a very strong correlation between the two scores; in fact, in this case, the largest modularity is achieved with three communities and the obtained communities are exactly the same as the ground

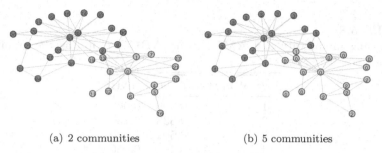

(a) 2 communities   (b) 5 communities

FIGURE 5.6
Clustering the Zachary graph with the **Girvan–Newman** algorithm. In (a), we show the result when we force 2 communities. In (b), we force 5 communities and use the clusters as labels while colouring the nodes with respect to the ground-truth.

truth (so that $AMI = 1$). In Table 5.8, we show the values for the modularity and **AMI** as the number of communities increases.

While 3 communities yield optimal value of the modularity function, there is little difference, for example, when one insists on having 4 communities. In Figure 5.9, we show the **ABCD** graph with the dendrogram cut to get 4 communities. In this example, we see a common behaviour of many graph clustering algorithms that recover many of the ground-truth communities but, say, one of them is split into two. In this case, the white colour nodes that

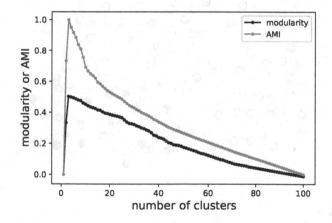

FIGURE 5.7
Clustering the **ABCD** graph with the **Girvan–Newman** algorithm. We compare **AMI** scores with the corresponding modularities.

TABLE 5.8
Comparing modularity and **AMI** for the **ABCD** graph as the number of communities increases.

# of parts	q	AMI
1	0.000	0.000
2	0.333	0.734
3	0.502	1.000
4	0.499	0.950
5	0.494	0.916

form a triangle are mistakenly put into a separate community. As usual, we direct the reader to the notebook for more details.

Quality Measures

In this chapter, we saw several measures of quality for clusterings, and we concluded that adjusted versions should be used in general. Let us illustrate here why this is recommended by doing the following simple experiment. We generated a graph G using the **ABCD** model with the following parameters: the number of nodes $n = 1,000$, the degree distribution exponent $\gamma = 2.5$ with

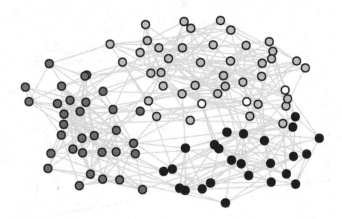

FIGURE 5.9
Clustering **ABCD** graph with 3 ground-truth communities when the algorithm is forced to find 4 communities. One of the ground truth community is split and its smaller component (white nodes) form a triangle.

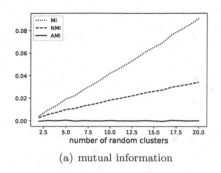
(a) mutual information

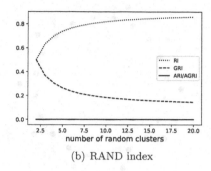
(b) RAND index

FIGURE 5.10
Comparing various quality measures for random partitions with varying number of parts.

degrees in the range [5, 50], the community size exponent $\tau = 1.5$ within the range [75, 150], and the mixing parameter $\xi = 0.1$. The obtained graph consists of 10 ground-truth communities. Next, we generated random partitions into s parts, for $2 \leq s \leq 20$. For each choice of s, we randomly generated 100 such partitions and computed different quality scores against the ground-truth partition.

Since those partitions are all random, assuming we wanted to compare the scores between partitions of different sizes, one should not expect much differences in the corresponding scores as we vary s. Moreover, it would be natural to expect that the scores should all be close to zero. We summarize our results in Figure 5.10. On the left plot, we compare the three variants of the mutual information scores: the original one (**MI**), normalized (**NMI**) and adjusted (**AMI**). We clearly see that **MI** and to a lesser extend **NMI** grow as a function of s, thus leading to the false conclusion that having 20 random clusters is better than having 2 random clusters. On the other hand, **AMI** is not influenced by the value of s and remains close to 0 for all values (as it should). In the right plot, we compare the RAND index (**RI**) and its adjusted counterpart (**ARI**). The difference here is even more pronounced, with large **RI** values that increase with s, while the **ARI** remains close to 0. We also compare the results with some recently proposed **graph-aware** measures, where the comparison is restricted to the pairs of nodes sharing an edge. We show results for the **graph-aware RAND index** (**GRI**) and its adjusted version (**AGRI**). Such measures tend to behave in a manner that is the opposite of their non graph-aware counterparts when the number of clusters varies, as we see on the plot for the non-adjusted measures. The conclusion is the same as before: we strongly recommend using adjusted measures such as **AMI**, **ARI** or **AGRI** when comparing results obtained by clustering algorithms.

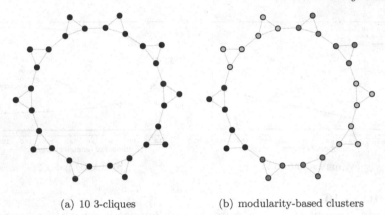

(a) 10 3-cliques   (b) modularity-based clusters

FIGURE 5.11
Example of a ring of 10 3-cliques (a) and resulting communities with modularity-based algorithm (b).

Modularity and Resolution Limit

Several graph clustering algorithms rely on optimizing the modularity function. As already mentioned earlier, this approach is known to have some drawbacks, one of them is well illustrated by considering rings of cliques. A **ring of cliques** is a graph that consists of ℓ complete subgraphs (cliques) of size s, with exactly one edge between cliques i and $i+1$ for $1 \leq i < \ell$, and one edge between clique ℓ and clique 1. Hence, there are $n = \ell s$ nodes and $m = \binom{s}{2}\ell + \ell$ edges in this graph. We illustrate an example of such graph in Figure 5.11(a), with $\ell = 10$ cliques of size $s = 3$ (triangles).

In Figure 5.12, we consider two algorithms that try to optimize the modularity function: **Louvain** and **CNM**. We investigated the resulting number of communities when considering the ring of cliques with $s = 3$ and $3 \leq \ell \leq 48$. We see that such algorithms often yield the number of communities that is much smaller than the number of cliques. This is due to the fact that the optimal value of the modularity is achieved by clumping contiguous cliques into the same community, as we illustrate in Figure 5.11(b). Using one of the ensemble methods such as **ECG**, one may improve the result as we see in Figure 5.12. The first step of the **ECG** algorithm consists of building several local clusterings by considering level-1 **Louvain** algorithm. This process tends to keep the edges within a clique in the same community, thus boosting their assigned weights for the last step of the **ECG**.

Nodes of Interest: k-hops vs. Community

Graph clustering is a useful unsupervised tool to study the structure of a graph and look for dense communities. However, there are other uses for such

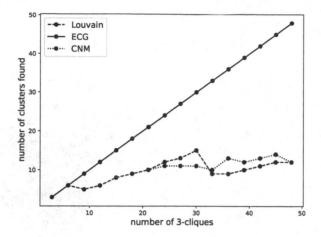

FIGURE 5.12
Modularity-based algorithms (**Louvain, CNM**) often merge cliques thus resulting in a smaller number of communities. One way to improve the result is to use an ensemble algorithm such as **ECG**.

algorithms and we illustrate one of them here. Our goal is to answer the following question. Given a large graph and some node of interest v, how do we sample the graph in order to restrict ourselves to the subgraph of nodes with the most interactions with node v? One possibility is to look at v's **ego-net**, that is, the subgraph obtained by considering only node v and its immediate neighbours, set $N(v) \cup \{v\}$. This can be extended to include nodes within distance k from node v. A disadvantage of this approach is that it can quickly yield large subgraphs and does not take edge density into account. Another approach one may want to consider is to perform graph clustering, and consider the cluster containing node v. With hierarchical clustering, one may additionally vary the level at which the dendrogram is cut to control the size of the subgraph containing v. With **ECG**, one may filter with respect to the induced edge weights (the number of votes) to control the size of the subgraph.

We illustrate this process using the airport graph which we already considered in Section 3.7. For simplicity, we ignore edge weights and directions. In Figure 5.13(a), we look at the ego-net for airport MQT (Marquette, MI) which has degree 11. In Figure 5.13(b), we look at the 2-hop neighbourhood of the same airport that already contains 221 nodes! The node of interest is shown in black and we always restrict ourselves to the 2-core of the subgraph. In Figure 5.13(c), we show the resulting community consisting of 48 nodes

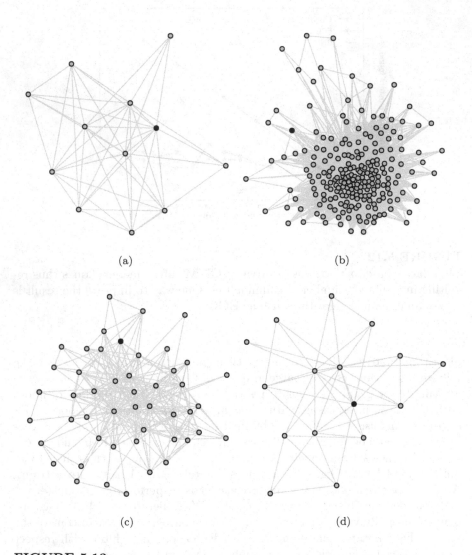

(a)   (b)

(c)   (d)

FIGURE 5.13
Various ways to look at subgraphs around the node corresponding to MQT (in black) in the airport graph. In (a) and (b), we consider respectively the 1- and the 2-hop ego-nets. In (c) we keep the **ECG** cluster containing MQT, and in (d) we filter this cluster keeping only edges with large number of **ECG** votes.

this node falls into when the **ECG** algorithm is applied. We may reduce this subgraph by only considering the edges with a large number of votes in the ensemble; in this case, we keep edges with **ECG** induced weight above 0.8, and we consider only the connected component containing the node of interest. As we see in Figure 5.13(d), as a result we find a tight subgraph consisting with 16 nodes, which contains 9 airports from MI, 6 from WI, and 1 from MO; of course, this may exclude some nodes from plot (a).

Comparing Clustering Algorithms

Let us illustrate a typical approach to compare graph clustering algorithms using benchmark graphs. We use the **ABCD** benchmark model with the following parameters: $n = 1,000$ nodes, node degrees between 10 and 50 with the power law exponent 2.5, and community sizes between 50 and 100 with the power law exponent 1.5. The mixing parameter ξ in the **ABCD** benchmark controls the level of noise: at $\xi = 0$ we have pure communities (no inter-community edges) and at the other extreme, $\xi = 1$, we have a random graph with no community structure.

In Figure 5.14, we show the resulting **AMI** values for 4 graph clustering algorithms: **Louvain, ECG, Infomap** and **Label Propagation**. We generated **ABCD** graphs with the above parameters and the mixing parameter in the following range: $0.1 \leq \xi \leq 0.8$. For each value of ξ, we independently generated 30 graphs. The curves in Figure 5.14 are obtained by taking the mean **AMI** values over those 30 graphs. As expected, all algorithms are very

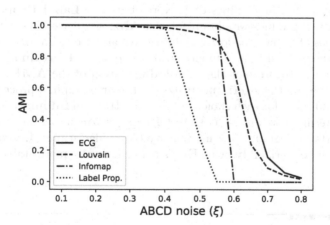

FIGURE 5.14
Comparing resulting communities vs. ground-truth for **ABCD** graphs with varying noise parameter ξ.

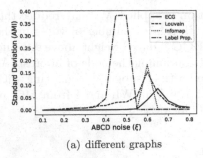 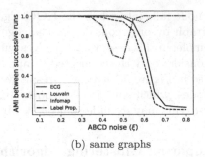

(a) different graphs   (b) same graphs

FIGURE 5.15
Standard deviation of the **AMI** values for the **ABCD** graphs: (a) over different graphs, and (b) over the same graph.

good at recovering the ground truth communities for small values of ξ, which correspond to dense communities with a small number of inter-community edges. As we increase ξ, we see that the performance of **Label Propagation** decays; the other algorithms seem to be able to tolerate a slightly higher level of noise, with **ECG** giving the best results in this case. Such studies can be done with various types of benchmark graphs and clustering algorithms.

Finally, let us have a look at the stability of those results from two different angles. In Figure 5.15(a), we plot the standard deviation of the **AMI** values we reported in Figure 5.14. From that plot we see a similar pattern for each algorithm: the standard deviation is relatively small until the results start to degrade (which happen at different values of ξ for each algorithm), and it goes down again for large values of ξ. Note that with **Label Propagation** and **Infomap**, the values we obtain for the **AMI** score are almost flat zero for large values of ξ, thus the low variability. We also see the impact of **ECG** in reducing the instability in comparison to the original **Louvain** algorithm.

In Figure 5.15(b), we plot the standard deviation of the **AMI** values we obtain when we run the algorithms twice on the same graph, a process which we repeated 30 times for each choice of ξ. We see that the **Infomap** algorithm is almost deterministic, and so is **Label Propagation** up to the noise level where it starts to degrade. The modularity-based algorithms (**Louvain** and **ECG**) show more variability, with **ECG** being generally more stable.

5.8   Practitioner's Corner

Graph clustering is a useful tool for mining graphs with a number of possible applications including finding communities, focusing on smaller, dense subgraphs around a given node (or a set of nodes) of interest, and visualization. Graph clustering is an unsupervised tool and, unless we are considering some synthetic benchmark model with embedded community structure (such as **ABCD**, **LFR** or **BTER**), there are no known "ground-truth" communities. To that effect, definitions such as weak or strong communities are useful to assess the value of the communities we obtain; this is important in unsupervised conditions since true dense communities may simply not exist in a given graph.

Hierarchical clustering algorithms allow us to consider several granularities for the communities which may be useful in practice. Other graph clustering algorithms, such as **Louvain** or **CNM**, often try to optimize the modularity function. The modularity is usually correlated with the ground-truth (at least for benchmark graphs) but the communities we find are often split or merged versions of the ground-truth.

One known issue with modularity based algorithms is the resolution limit problem; for example, merging natural and strong communities such as cliques sometimes leads to higher modularity. One way to reduce this problem is to use an ensemble learning approach such as **ECG** that also allows to quantify the strength of each edge of being "within" a community. Finally, let us stress it again that the ground-truth is usually not available for real graphs, so strong conclusions should generally be avoided and graph clustering should be used mainly as an EDA tool.

5.9 Problems

In this section we present a collection of potential practical problems for the reader to attempt.

1. Run various clustering algorithms (**ECG**, **Louvain**, **Infomap**, **Label Propagation**, **Girvan-Newman**, and **CNM**) for the karate club graph. For each algorithm tested, compare the partition found by the algorithm with the ground-truth (two communities) by computing the **AMI** score.

2. Run various clustering algorithms (**ECG**, **Louvain**, **Infomap**, **Label Propagation**, **Girvan-Newman** and **CNM**; note that **CNM** might be too slow but give it a try anyways) on the GitHub ml graph. Which algorithms produce similar results? In order to answer this question, for each pair of algorithms, find the **AMI** score between the two results.

3. Re-do the experiment used to generate Figure 5.14 but additionally include the **Girvan-Newman** algorithm. That is, for each graph tested, run the **Girvan-Newman** algorithm, select the partition with the highest modularity, and use it for comparison to the

ground-truth (**AMI**). (If the experiment is slow, then generate only, say, 3 graphs instead of 30.)

4. Similarly to the experiment used to generate Figure 5.14, independently generate 30 copies of the **ABCD** graph with varying noise parameter ξ's, and run the **Louvain** algorithm to obtain different partitions that represent communities. For each partition compute the average ratio $\deg^{int}(v)/\deg^{out}(v)$ over all nodes in the graph (see the definition of strong community). Plot the average of the average ratios for all partitions over 30 graphs. Independently, for each partition compute the fraction of communities forming a weak communities, and plot the average fraction over 30 graphs.

5. Take the **ABCD** graph we used to test quality measures (see Figure 5.10). Check node roles. Compute how many nodes we have in each family (recall that there are 4 families of non-hubs and 3 families of hubs). Plot the $(z(v), p(v))$ scores for all nodes as we did in Figure 5.3 for the karate club graph.

6. Compare time complexities of various clustering algorithms (**ECG**, **Louvain**, **Infomap**, **Label Propagation**, **Girvan-Newman**, **CNM**) using an **ABCD** synthetic graph with different number of nodes, say, $n = 100, 200, 400, 800, 1600, \ldots$. Which algorithm is the slowest, which one is the fastest?

7. Re-do the experiment with the ring of cliques (Figure 5.12) but instead of using $s = 3$ (triangles), check $s = 5$ (K_5) and $s = 7$ (K_7).

5.10   Recommended Supplementary Reading

- F. Radicchi, C. Castellano, F. Cecconi, V. Loreto, D. Parisi. "Defining and identifying communities in networks". *PNAS* (2004), 101:2658–2663. (Communities)

- R. Guimerá, L.A.N. Amaral. "Cartography of complex networks: modules and universal roles". *J. Stat. Mech.-Theory E.* **2005** (2005), P02001. (Node Roles)

- P.W. Holland, K.B. Laskey, S. Leinhardt, "Stochastic blockmodels: First steps". *Social Networks* **5** (1983) 109–137. (Stochastic Block Model)

- A. Lancichinetti, S. Fortunato, F. Radicchi. "Benchmark graphs for testing community detection algorithms". *Phys. Rev. E*, **78** (2008), 046110. (LFR Model)

- B. Kamiński, P. Prałat, F. Théberge. "Artificial Benchmark for Community Detection (ABCD) — Fast Random Graph Model with Community Structure". *Network Science* 9(2) (2021), 153–178. (ABCD Model)

- MEJ Newman, M. Girvan. "Finding and evaluating community structure in networks". *Phys. Rev. E.* 2004; 69: 026–113. (Graph Modularity)

- V.D. Blondel, J.-L. Guillaume, R. Lambiotte, E. Lefebvre. "Fast unfolding of communities in large networks". *Journal of Statistical Mechanics: Theory and Experiment* (2008), P10008. (Louvain Algorithm)

- V. Poulin, F. Théberge. "Ensemble Clustering for Graphs: Comparison and Applications". *Applied Network Science* 4 (2019), 51. (ECG Algorithm)

- S. Fortunato, M. Barthelemy. "Resolution limit in community detection". *Proc. Natl. Acad. Sci. USA.* 2007: 104: 36–41. (Resolution Limit)

- E. Ravasz, A.L. Somera, D.A. Mongru, Z.N. Oltvai, A.-L. Barabási. "Hierarchical organization of modularity in metabolic networks". *Science* **297** (2002), 1551–1555. (Ravasz Algorithm)

- M. Girvan, M.E.J. Newman. "Community structure in social and biological networks". *P. Natl. Acad. Sci. USA* **99** (2002), 7821–7826. (Girvan–Newman Algorithm)

- U.N. Raghavan, R. Albert, S. Kumara. "Near linear time algorithm to detect community structure in large-scale networks". *Phys. Rev. E* **76** (2007), 036106. (Label Propagation Algorithm)

- M. Fiedler. "A Property of Eigenvectors of Nonnegative Symmetric Matrices and its Application to Graph Theory". *Czechoslovak Mathematical Journal* **25(4)** (1975) 619–633. (Spectral Bisection Algorithm)

- M. Rosvall, D. Axellson, C.T. Bergstrom. "The map equation". *The European Physical Journal Special Topics* **178** (2009), 13–23. (Infomap)

- V. Poulin, F. Théberge. "Comparing Graph Clusterings: Set partition measures vs. Graph-aware measures". *IEEE Transactions on Pattern Analysis and Machine Intelligence*, in press. (Graph-aware measures for comparing partitions)

- G.M. Slota, J. Berry, S.D. Hammond, S. Olivier, C. Phillips, S. Rajamanickam. "Scalable generation of graphs for benchmarking hpc community-detection algorithms". *IEEE international conference for high performance computing, networking, storage and analysis (SC)* (2019). (BTER)

- C.L. Staudt, M. Hamann, A. Gutfraind, I, Safro, H. Meyerhenke. "Generating realistic scaled complex networks". *Applied network science* **2(36)** (2017), 1–29. (ReCoN)

The Zachary Karate Club dataset introduced in this chapter originates from:

- W.W. Zachary, "An Information Flow Model for Conflict and Fission in Small Groups" *Journal of Anthropological Research*, (33) 452–473 (1977).

6

Graph Embeddings

6.1 Introduction

The goal of many machine learning applications is to make predictions or discover new patterns using graph-structured data as feature information. For example, one might want to better understand a person's role within the collaboration network, similarity between users interacting on Amazon or Yelp, a protein's behaviour in a biological interaction network, or make recommendations to users of some social media platform.

In order to extract useful structural information from graphs, one might want to try embedding it in a geometric space by assigning coordinates to each node such that nearby nodes are more likely to share an edge than those far from each other or are similar. In particular, in the case of link prediction, a good embedding should have the property that most of the networks edges can be predicted from the coordinates of the nodes. On the other hand, in the case of node classification, one might want to include information about the global position of a node in the graph or the structure of the node's local graph neighbourhood. Other applications might require different properties to be preserved. Hence, unfortunately, in the absence of a general-purpose representation for graphs, very often graph embedding requires domain experts to craft features or to use specialized feature selection algorithms.

The very first graph embedding techniques from the early 2000's were designed as dimension reduction methods for non-relational data. Such data, possibly living on a manifold, can be represented as a graph in several ways. For example, each data point can be mapped to a node in a graph, and edges are built via some affinity-graph transformation such as k-nearest neighbours (for a given node v corresponding to a data point, one adds an edge between v and its k-nearest neighbours) or ϵ-ball (simply one adds edges between points at distance at most ϵ). Since then, embedding algorithms evolved and now are commonly used for datasets already represented as a graph. In order to produce an embedding, one needs to assume that some proximity measure between the graph's nodes is provided. With a clear objective at hand, there are still various possible techniques and approaches that one may try to utilize to generate the desired embedding. Hence, it is not uncommon that the user is left with several embeddings of its nodes in some multidimensional spaces (possibly in different dimensions) and tries to decide which one should be used. Fortunately, there are some unsupervised tools that may be used to evaluate these embeddings.

This chapter is structured as follows. We first introduce a few proximity measures between nodes that a good embedding algorithm might want to preserve (Section 6.2). Then, we highlight a few techniques to achieve it, including algorithms that use linear algebra, are based on the theory of random walks, and the ones which use deep learning (Section 6.3). Because of the abundance of various algorithms, it is important to be able to select the best embedding from a large collection of embeddings to choose from (or at least, to ignore the bad ones). We present one unsupervised framework to benchmark embeddings (Section 6.4). There are many important applications of node embeddings that we discuss next (Section 6.5). We tried to focus on the most important aspects, but clearly there are many other interesting directions in this area that the readers might want to explore (Section 6.6). As usual, we finish the chapter with experiments (Section 6.7) and provide some tips for practitioners (Section 6.8).

Finally, let us mention that in this chapter we concentrate on embedding nodes of a given graph. Embedding graphs that belong to some family of graphs is separately discussed in Chapter 9.

6.2   Problem Formalization

Let $G = (V, E)$ be a weighted graph on the set of nodes $V = \{v_1, v_2, \ldots, v_n\}$. For simplicity, we first deal with undirected graphs before briefly commenting on how one can deal with directed graphs. An **embedding** is a function $\mathcal{E} \colon V \to \mathbb{R}^k$, where k is much smaller than n. In other words, the embedding represents each node as a low-dimensional feature vector. The goal of the function $\mathcal{E}$ is not only to decrease the dimension but to also preserve pairwise proximity between nodes as best as possible. In this section, we introduce a few natural proximity measures to be preserved in the embedded space. Each of them produces a matrix $\mathbf{S} = (s(v_i, v_j))_{i,j \in [n]}$, where $s(v_i, v_j)$ measures the proximity between nodes v_i and v_j as a non-negative real number. Since the distance between two points in $\mathbb{R}^k$ is symmetric, it is desirable for matrix $\mathbf{S}$ to be symmetric (though not all proximity measures guarantee this property). If $\mathbf{S}$ is not symmetric, then it is left for the embedding algorithm to interpret this and come up with a good function $\mathcal{E}$. Note also that the diagonal of $\mathbf{S}$ has a special role. Various methods produce different values on the diagonal but, in general, these values should be ignored by embedding algorithms.

The first-order proximity is the local pairwise similarity between nodes connected by an edge. Two nodes are simply more similar if they are connected by an edge with larger weight.

Let $G = (V, E)$ be any graph on the set of nodes $V = \{v_1, v_2, \ldots, v_n\}$. The **first-order proximity** $s_1(v_i, v_j)$ between node v_i and node v_j is the weight of the edge $v_i v_j$, that is, $s_1(v_i, v_j) = a(v_i, v_j)$, where $\mathbf{A} = (a(v_i, v_j))_{i,j \in [n]}$ is the adjacency matrix.

The second-order proximity compares the similarity between the neighbourhoods of the nodes. The more similar two nodes' neighbourhoods are, the larger the second-order proximity value between them is.

Let $G = (V, E)$ be any graph on the set of nodes $V = \{v_1, v_2, \ldots, v_n\}$. The **second-order proximity** $s_2(v_i, v_j)$ between node v_i and node v_j is a similarity between v_i's neighbourhood

$$s_1(v_i) = (s_1(v_i, v_1), s_1(v_i, v_2), \ldots, s_1(v_i, v_n))$$

and v_j's neighbourhood

$$s_1(v_j) = (s_1(v_j, v_1), s_1(v_j, v_2), \ldots, s_1(v_j, v_n)).$$

This similarity can be measured, for example, using the **cosine similarity** metric, which is a standard measure of similarity between two non-zero vectors. It is defined as the cosine of the angle between them which is the same as the inner product of the corresponding normalized vectors, that is,

$$s_2(v_i, v_j) = \frac{\sum_{\ell=1}^{n} s_1(v_i, v_\ell)\, s_1(v_j, v_\ell)}{\sqrt{\sum_{\ell=1}^{n} s_1(v_i, v_\ell)^2}\, \sqrt{\sum_{\ell=1}^{n} s_1(v_j, v_\ell)^2}}.$$

These definitions naturally generalize to higher-order proximities. For any $k \in \mathbb{N} \setminus \{1, 2\}$, the **$k$th-order proximity** $s_k(v_i, v_j)$ between node v_i and v_j is the similarity between v_i's $(k-1)$st neighbourhood $s_{k-1}(v_i)$ and v_j's $(k-1)$st neighbourhood $s_{k-1}(v_j)$.

There are many other possible ways to measure proximity between nodes. Alternatively, one may want to use some centrality measure such as the Katz Index, Personalized PageRank, Common Neighbours, or Adamic Adar. The first definition builds on the Katz centrality measure discussed in Section 3.2. It tries to capture the relative influence of a node within a network by considering walks of any length but penalizes longer walks by introducing the attenuation factor α.

Let $G = (V, E)$ be any graph on n nodes. Fix any α such that $0 < \alpha < \min\{1, 1/|\lambda|\}$, where λ is the leading eigenvalue of adjacency matrix $\mathbf{A}$. The **Katz Index** $\mathbf{S}_\alpha^{\mathrm{Katz}}$ is defined as follows:

$$\mathbf{S}_\alpha^{\mathrm{Katz}} = \sum_{i=1}^{\infty} (\alpha \cdot \mathbf{A})^i = (\mathbf{I}/\alpha - \mathbf{A})^{-1} \mathbf{A}.$$

The next centrality measure, personalized PageRank, is inspired by the well-known PageRank centrality measure. Personalized PageRank depends on two parameters: the **jumping constant** α and the **seed** node $s \in V$. In this variant, which is commonly used by web search engines to find the most relevant pages to a given request, a random walk continues with probability α and goes back to the seed node s with probability $1 - \alpha$. The original definition of PageRank centrality measure is the special case of its personalized counterpart where each time the teleportation takes place, a seed is uniformly sampled from V. The personalized PageRank proximity matrix is then constructed by combining the n columns, each of length n, generated by considering all nodes as seeds for the personalized PageRank procedure described above.

Let $G = (V, E)$ be any graph on n nodes. Fix any $0 < \alpha < 1$. The **Personalized PageRank** $\mathbf{S}_\alpha^{\mathrm{PPR}}$ is defined as follows:

$$\mathbf{S}_\alpha^{\mathrm{PPR}} = (1 - \alpha) \left(\mathbf{I} - \alpha \hat{\mathbf{A}}^T \right)^{-1},$$

where matrix $\hat{\mathbf{A}}$ is defined in (3.6).

Let us mention that $\mathbf{S}_\alpha^{\mathrm{PPR}}$ is usually not symmetric. We will encounter the same problem for directed graphs. As mentioned above, this can be dealt with at the level of the embedding algorithm. Alternatively, one may apply some natural transformation to make the matrix symmetric. We will come back to this at the end of this section.

In the next definition, we simply count the number of nodes that have both v_i and v_j as their neighbours. As mentioned earlier (see, for example, equation (3.4)), the element (v_i, v_j) of $\mathbf{A}^k$ is equal to the number of walks of length k from node v_i to node v_j. Using this observation, we get the following definition that is closely related to the **topological overlap matrix** we discussed in Section 5.5.

Let $G = (V, E)$ be any graph on n nodes. Given adjacency matrix $\mathbf{A}$, the **Common Neighbours** $\mathbf{S}^{\mathrm{CN}}$ is defined as follows:

$$\mathbf{S}^{\mathrm{CN}} = \mathbf{A}^2.$$

The next proximity measure is a variant of the common neighbours measure. This time given two nodes, the proximity measure between them is the sum of the reciprocals of the logarithms of the degrees of their common neighbours. As a result, the fact that two nodes are common neighbours of some nodes that have very large neighbourhoods is less significant.

Let $G = (V, E)$ be any graph on the set of nodes $V = \{v_1, v_2, \ldots, v_n\}$. The **Adamic-Adar** $\mathbf{S}^{\mathrm{AA}}$ is defined as follows:

$$s^{AA}(v_i, v_j) = \sum_{v_k \in N(v_i) \cap N(v_j)} \frac{1}{\ln(\deg(v_k))},$$

for $i \neq j$ and $s^{AA}(v_i, v_i) = 0$.

Let us now turn our attention to directed graphs for which matrix $\mathbf{A}$ might not be symmetric. There are various methods proposed within the literature to handle this situation. The simplest ones are either to ignore it and compute $\mathbf{S}$ using the original adjacency matrix $\mathbf{A}$, or to consider $(\mathbf{A} + \mathbf{A}^T)/2$ instead, that is, to transform the graph to its undirected counterpart by replacing two directed edges between v_i and v_j (in both directions) with a single undirected edge which has a weight equal to the average weight of the two corresponding directed edges. However, even after implementing such an averaging operation, some proximity measures might still produce an asymmetric proximity matrix $\mathbf{S}$. If this happens, then one may leave it for the embedding algorithm to deal with. Some algorithms such as **Local Linear Embedding (LLE)** and **High Order Proximity preserving Embedding (HOPE)**, which are mentioned in the next section, are explicitly designed to handle this situation. Alternatively, one may simply pass a symmetric matrix $(\mathbf{S} + \mathbf{S}^T)/2$ to an embedding algorithm in lieu of $\mathbf{S}$.

The last proximity measure that we would like to mention, **SimRank**, can be used to explicitly compute $\mathbf{S}$ for directed graphs. The idea behind it is similar to other second-order proximity measures but it does not require the neighbouring nodes of the considered pair of nodes to be identical; it is enough if they are similar.

Let $D = (V, E)$ be any directed graph on the set of nodes $V = \{v_1, v_2, \ldots, v_n\}$. The **SimRank S^{SR}** is defined as follows: $s^{SR}(v_i, v_i) = 1$, $s^{SR}(v_i, v_j) = 0$ if in-degree of v_i is equal to 0 or in-degree of v_j is equal to 0. For the remaining pairs of nodes, the following recurrence relation should be satisfied:

$$s^{SR}(v_i, v_j) = \frac{C}{\deg^{in}(v_i)\,\deg^{in}(v_j)} \cdot \sum_{k=1}^{n}\sum_{\ell=1}^{n} a(v_k, v_i) \cdot a(v_\ell, v_j) \cdot s^{SR}(v_k, v_\ell),$$

where $C \in [0, 1]$ is a universal normalizing constant.

It is easy to see that the **SimRank** is uniquely determined for any normalizing constant $C \in [0, 1]$. The simplest method of approximating $\mathbf{S}$ is to start with a diagonal matrix $\mathbf{I}$ and iteratively apply the above recurrence relation to it. As a consequence, the entries of $\mathbf{S}$ matrix belong to the interval $[0, 1]$ and larger similarity scores correspond to pairs of nodes whose in-neighbours are *on average* also similar.

6.3   Techniques

In this section, we discuss various methods for embedding the nodes of a graph in a vector space. The goal is not to create an exhaustive list of embedding algorithms with all details of their implementations carefully explained, but rather to build an understanding of possible approaches and techniques to construct the desired embeddings.

Linear Algebra Algorithms

Local Linear Embedding (LLE) is a simple instance of matrix factorization which uses the (weighted) adjacency matrix $\mathbf{A}$ whose elements $a(v_i, v_j)$ represent the weight of an edge $v_i v_j$. This algorithm works for both undirected and directed graphs so, in order to cover both scenarios, we do *not* assume that $\mathbf{A}$ is symmetric but it might be. Recall that for any $v_i \in V$, $N^{out}(v_i) \subseteq V$ denotes the set of out-neighbours of v_i. In what follows, we assume that for all nodes v_i the set $N^{out}(v_i)$ is non-empty. Let $e_i = \mathcal{E}(v_i)$ be the embedding of node v_i. For convenience, let $\mathbf{E}$ be the $k \times n$ matrix consisting of vectors e_i that form the columns of $\mathbf{E}$. It will be useful to denote by $\hat{a}(v_i, v_j)$ elements of matrix $\hat{\mathbf{A}}$ defined in (3.6), that is,

$$\hat{a}(u, v) = a(u, v)/\deg^{out}(u)$$

(recall that $\deg^{out}(u) > 0$).

With **LLE**, our goal is to locally and linearly approximate e_i by a weighted sum of the embeddings of its neighbours, that is, e_i should be close to

$$\sum_{v_j \in N^{out}(v_i)} \hat{a}(v_i, v_j)e_j = \sum_{j=1}^{n} \hat{a}(v_i, v_j)e_j.$$

We may rewrite it conveniently in a matrix form as a task of minimizing the following optimization function:

$$\Phi(\mathbf{E}) = \left\| (\mathbf{I} - \hat{\mathbf{A}})\mathbf{E}^T \right\|_F,$$

where $\|\cdot\|_F$ is the **Frobenius norm** that is a natural extension of the Euclidean norm to matrices in $\mathbb{R}^{n \times m}$: for any $\mathbf{B} = (b_{i,j})_{i \in [m], j \in [n]} \in \mathbb{R}^{n \times m}$,

$$\|\mathbf{B}\|_F = \sqrt{\sum_{i=1}^{m} \sum_{j=1}^{n} b_{i,j}^2}.$$

Since clearly $\|\mathbf{B}\|_F = \left\|\mathbf{B}^T\right\|_F$, we may alternatively express the objective function as

$$\Phi(\mathbf{E}) = \left\| ((\mathbf{I} - \hat{\mathbf{A}})\mathbf{E}^T)^T \right\|_F = \left\| \mathbf{E}(\mathbf{I} - \hat{\mathbf{A}})^T \right\|_F = \left\| \mathbf{E}(\mathbf{I} - \hat{\mathbf{A}}^T) \right\|_F. \qquad (6.1)$$

We immediately note that, for example, $\mathbf{E} = \mathbf{0}_{k \times n}$ is a valid solution to unconstrained problems of this sort. Therefore, it is natural to add some additional conditions that will allow us to find a non-degenerate embedding matrix $\mathbf{E}$. First, note that $\hat{\mathbf{A}}\mathbf{1} = \mathbf{1}$ by construction of matrix $\hat{\mathbf{A}}$. Therefore, $\Phi(\mathbf{E}) = \Phi(\mathbf{E} + \mathbf{x}\mathbf{1}^T)$ for any k-dimensional vector $\mathbf{x}$. Indeed, in order to see this observe that

$$(\mathbf{I} - \hat{\mathbf{A}})\mathbf{1}\mathbf{x}^T = \mathbf{1}\mathbf{x}^T - \hat{\mathbf{A}}\mathbf{1}\mathbf{x}^T = \mathbf{1}\mathbf{x}^T - \mathbf{1}\mathbf{x}^T = \mathbf{0}.$$

Therefore, we assume that $\mathbf{E}\mathbf{1} = \mathbf{0}$, that is, we insist that rows of $\mathbf{E}$ are centred around the origin. In order to further constrain the admissible solution space, we require that rows of $\mathbf{E}$ have equal norms and are mutually orthogonal. This condition is typically expressed as $\mathbf{E}\mathbf{E}^T/n = \mathbf{I}$, which together with $\mathbf{E}\mathbf{1} = \mathbf{0}$, means that rows of $\mathbf{E}$ have a unit covariance. The interpretation is that if we know one dimension of embedding $\mathbf{E}$, it provides no information about the missing dimensions; the linear prediction of the missing values for any node is equal to 0. Given these conditions, the solution is still identified only up to the rotation. Indeed, for any rotation matrix $\mathbf{R}$ (that is, $\mathbf{R}^T = \mathbf{R}^{-1}$ and $\det(\mathbf{R}) = 1$), we get that

$$\mathbf{R}\mathbf{E}\mathbf{1} = \mathbf{R}\mathbf{0} = \mathbf{0}$$
$$\mathbf{R}\mathbf{E}(\mathbf{R}\mathbf{E})^T/n = \mathbf{R}\mathbf{E}\mathbf{E}^T\mathbf{R}^T/n = \mathbf{R}\mathbf{I}\mathbf{R}^T = \mathbf{R}\mathbf{R}^T = \mathbf{R}\mathbf{R}^{-1} = \mathbf{I},$$

and, since it is known that the Frobenius norm is invariant with regards to rotations, we get that

$$\Phi(\mathbf{RE}) = \left\| \mathbf{RE}(\mathbf{I} - \hat{\mathbf{A}}^T) \right\|_F = \left\| \mathbf{E}(\mathbf{I} - \hat{\mathbf{A}}^T) \right\|_F = \Phi(\mathbf{E});$$

in the formulas above we used the alternative definition of Φ where the embedding matrix is the first term in the product under the Frobenius norm—see (6.1).

In order to minimize the objective function $\Phi(\mathbf{E})$, it is convenient to rewrite it as $\mathrm{tr}(\mathbf{E}(\mathbf{I} - \hat{\mathbf{A}})^T(\mathbf{I} - \hat{\mathbf{A}})\mathbf{E}^T)$, where the trace $\mathrm{tr}(\mathbf{C})$ of a square matrix $\mathbf{C}$ is defined to be the sum of elements on the main diagonal of $\mathbf{C}$. Now, since $\mathbf{M} = (\mathbf{I} - \hat{\mathbf{A}})^T(\mathbf{I} - \hat{\mathbf{A}})$ is symmetric and semi-positive definite ($\mathbf{z}^T M \mathbf{z} \geq 0$ for any vector $\mathbf{z} \in \mathbb{R}^n$) we know that $\mathbf{M}$ has non-negative eigenvalues. Note then that 0 is its smallest eigenvalue and is associated with an eigenvector $\mathbf{1}$ which we ruled out earlier. Since $\mathbf{M}$ is symmetric, all of its eigenvectors are orthogonal. It follows that, with the exception of the disallowed $\mathbf{1}$ vector, they meet the condition of being centred in the origin if we consider their inner product with $\mathbf{1}$ eigenvector, which is equal to $\mathbf{0}$. A final observation is that $\mathrm{tr}(\mathbf{E}(\mathbf{I} - \hat{\mathbf{A}})^T(\mathbf{I} - \hat{\mathbf{A}})\mathbf{E}^T)$ is minimized for $\mathbf{E}$ whose rows are the k eigenvectors corresponding to the smallest eigenvalues of $(\mathbf{I} - \hat{\mathbf{A}})^T(\mathbf{I} - \hat{\mathbf{A}})$, excluding the smallest eigenvalue that is equal to 0 (which produces a non-centred eigenvector as discussed above). Note that, when normalized, these eigenvectors meet the desired condition $\mathbf{EE}^T/n = \mathbf{I}$, as they are orthogonal.

In summary, in order to minimize $\Phi(\mathbf{E})$ one needs to find the $k+1$ eigenvectors of the $n \times n$ matrix $(\mathbf{I} - \hat{\mathbf{A}})^T(\mathbf{I} - \hat{\mathbf{A}})$ corresponding to the smallest $k+1$ eigenvalues and discard the eigenvector $\mathbf{1}$ that corresponds to the smallest eigenvalue that is equal to 0 (recall that k is much smaller than n). From a numerical perspective, let us mention that there exist algorithms that allow one to find these eigenvectors without evaluating the product $(\mathbf{I} - \hat{\mathbf{A}})^T(\mathbf{I} - \hat{\mathbf{A}})$. Finally, when we take an interpretive perspective, we observe that **LLE** tries to find an embedding such that the vector e_i corresponding to node v_i is close in the embedded space to vectors that are associated with nodes that are adjacent to node v_i—this brings to mind the first-order proximity mentioned in Section 6.2.

The **Laplacian Eigenmaps** algorithm (**LEM**) uses the graph Laplacian matrix $\mathbf{L} = \mathbf{D} - \mathbf{A}$, which assumes that $\mathbf{A}$ is symmetric and $\mathbf{D} = (d(v_i, v_j))_{i,j \in [n]} = \mathrm{diag}(\mathbf{A1})$ is the diagonal matrix with $d(v_i, v_i) = \deg(v_i) = \sum_{j=1}^{n} a(v_i, v_j)$. This time our goal is to minimize the following optimization function:

$$\Phi(\mathbf{E}) = \sum_{(v_i, v_j) \in V^2} \|e_i - e_j\|^2 \, a(v_i, v_j) = \mathrm{tr}(\mathbf{ELE}^T),$$

that is, points that are connected by heavily weighted edges should be close to each other in the embedded space.

Similarly as for **LLE**, we see that **1** is a solution to this equation that we rule out. As before, we insist that $\mathbf{ED1} = \mathbf{0}$ and $\mathbf{EDE}^T = \mathbf{I}$ with the same interpretation as for **LLE**; the only difference is that this time we also weigh the nodes by their degrees. Denoting $\mathbf{E}' = \mathbf{ED}^{1/2}$, we see that our problem is equivalent to minimizing $\mathrm{tr}(\mathbf{E}'\mathbf{D}^{-1/2}\mathbf{LD}^{-1/2}\mathbf{E}'^T)$ subject to $\mathbf{E}'\mathbf{E}'^T = \mathbf{I}$. Therefore, we may apply the same sort of reasoning that we applied to **LLE** to conclude that the problem can be solved by computing the eigenvectors of the normalized Laplacian matrix $\mathbf{D}^{-1/2}\mathbf{LD}^{-1/2}$ corresponding to its smallest $k + 1$ eigenvalues (excluding the eigenvalue equal to 0, which we ruled out above; note that the 0 eigenvalue always exists and has multiplicity of one, provided the graph is connected).

From an interpretive perspective, similarly to **LLE**, **LEM** tries to find an embedding such that the vector e_i representing node v_i is close in the embedded space to vectors that represent adjacent nodes. In both methods, the goal is to locally and linearly approximate the location e_i of node v_i by a weighted sum over all locations of its outgoing neighbours. The difference is that in **LLE** the weights are normalized to one for each node whereas in **LEM** nodes that have many and/or heavy adjacent edges get more weight in the objective function. Finally, note that in this procedure one can replace $\mathbf{A}$ by any proximity measure $\mathbf{S}$ as long as it is symmetric. As a result, one can easily apply these techniques to some other proximity measure instead of the first-order proximity.

The last approach we would like to highlight that uses linear algebra to find a suitable embedding minimizes the following loss function

$$\Phi(\mathbf{E}) = \left\| \mathbf{S} - \mathbf{E}^T\mathbf{E} \right\|_F,$$

where $\mathbf{S}$ is some proximity measure matrix and, as usual, the columns of matrix $\mathbf{E}$ are the embeddings. This amounts to keeping the inner product of the embedding vectors of two given nodes as close as possible to the similarity between the nodes in the graph. This is a general approach in which any proximity measure, including these that we discussed in Section 6.2, can be used. In particular, let us mention about the **HOPE** algorithm which is an interesting instance of this approach aimed at embedding nodes in directed graphs. For every node v_i, we define two embeddings, $e_{s,i}$ and $e_{t,i}$, the source and, respectively, the target embedding. Let $\mathbf{E}_s$ and $\mathbf{E}_t$ be the corresponding matrices of the source and the target embeddings. The loss function for **HOPE** for a given similarity matrix $\mathbf{S}$, is defined as follows:

$$\Phi(\mathbf{E}_s, \mathbf{E}_t) = \left\| \mathbf{S} - \mathbf{E}_s^T\mathbf{E}_t \right\|_F.$$

Depending on the application at hand, one might later utilize $\mathbf{E}_s$ or $\mathbf{E}_t$ for further analysis. Alternatively, one may vertically concatenate both embeddings to capture the differences between the roles that nodes play as sources and targets within the graph (at the cost of increasing the dimensionality of the resulting embedding).

As usual, let us make some general comments about this approach. Implementations of algorithms from this class of embeddings involve performing a singular value decomposition of matrix $\mathbf{S}$. From the perspective of interpreting the embedding we can make the observation that if two nodes v_i and v_j are embedded close to each other, then they have similar values in the corresponding columns of matrix $\mathbf{S}$.

Finally, let us note that this setting is quite flexible. In particular, one can easily adjust the objective function to satisfy any desired property that can be specific to an application at hand. For instance, the $\left\| \mathbf{S} - \mathbf{E}^T \mathbf{E} \right\|_F$ is sensitive to the values on the diagonal of matrix $\mathbf{S}$, and one might want to add a correction in order to avoid this potentially undesired behaviour. Similarly, regularization of matrix $\mathbf{E}$ is often considered and results in a higher quality outcome.

Algorithms Based on Random Walks

Random walk based node embedding methods are derived from the **Word2Vec** algorithm for word embedding commonly used in **Natural Language Processing (NLP)**. **Word2Vec** is based on the assumption that "words are known by the company they keep." For a given word, an embedding is produced by forming context windows from sentences containing the word. A context window for a given word typically consists of (up to) ℓ preceding and ℓ following words in a sentence. We assume that we have a set of words W and each word is uniquely associated with a number in $[w]$, where $w = |W|$.

The **Word2Vec** implementation we consider uses a model known as **Skip-Gram** to learn the embedding. For example, consider a context window of size 5 (that is, we take $\ell = 2$) and the sentence: "Graph *embedding maps* **nodes** *to vector* space." The model is trained to predict the words in *italics* given the word "**nodes**" as input. Similarly to **HOPE** which we discussed earlier, **SkipGram** associates two embedding vectors with each word. Let $\mathbf{E}_s$ and $\mathbf{E}_t$ be the respective matrices of all "source" and "target" embeddings. These two matrices try to capture relationships between the input words and the words that they aim to predict and vice versa. In both matrices, the i-th columns (denoted respectively as $e_{s,i}$ and $e_{t,i}$) are associated with word $i \in [w]$. As a result, these matrices have dimension $k \times w$, where k is the desired dimension of the embedding.

Given a word $i \in [w]$, the probability $p_{i,j}$ that we see word $j \in [w]$ in its neighbourhood is approximated by the **softmax function** $q_{i,j}$ as follows:

$$q_{i,j} = \frac{\exp(e_{t,j}^T e_{s,i})}{\sum_{\ell=1}^{|W|} \exp(e_{t,\ell}^T e_{s,i})}. \tag{6.2}$$

The model is trained using maximum likelihood estimation.

In order to adapt such word-based techniques to graphs, we need to find analogous notions of words, sentences, and nearby words that apply to graphs, and adjust the objective function that needs to be optimized. The main challenge is that for typical empirical graphs the number of nodes is much larger than the size of the corpus of words in the typical setting wherein NLP techniques are applied. In order to highlight techniques used in this area, we concentrate on two selected algorithms, **Deep Walk** and **Node2Vec**, and describe their respective approaches to this task.

The common and general idea is as follows. The words are simply the nodes of a graph, and we generate sentences (sequences of nodes) via random walks on a graph. The exact procedure of how one performs such random walks differs for the two algorithms. We perform the random walk for a pre-defined number of steps and then "extract" the sentences from it. For example, if we consider a tiny directed graph in Figure 6.1, then some possible walks of length 4 are: $A \to B \to C \to B$, $B \to D \to C \to B$, and $C \to B \to D \to C$.

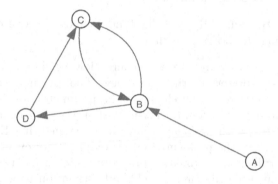

FIGURE 6.1
Random walks on directed graphs.

In the **Deep Walk** algorithm, the family of walks is sampled by performing random walks on G, typically between 32 and 64 per node, and for some fixed length. The walks are then used as sentences. Depending on the size of the context window, it is therefore possible to explore node's neighbourhood beyond its first- and second- order proximity. Now, for each node v_i, the algorithm tries to find an embedding e_i of v_i that maximizes the approximated likelihood of observing the nodes in the context windows obtained from the generated walks, assuming independence of observations.

Let us mention briefly some important complexity challenges that affect the algorithm. In practice, it is not possible to use the model given in (6.2) to make the predictions since, as mentioned above, the number of nodes is usually much larger than the number of words in the typical **NLP** application. The denominator in (6.2) is very expensive to compute and potentially numerically unstable. Therefore, the procedure uses a predictive model called

hierarchical softmax. In this approach, each node in the graph is represented as a leaf in a binary tree, and a binary classifier is fitted at each split of this binary tree. Each of these classifiers takes the embedding of source nodes as its input and then approximates the probability of observing a target node as a product of the approximated probabilities produced by the classifiers from root to leaf in the binary tree. Note that in this way we only need to evaluate $\Theta(\ln n)$ classifiers (where n is the number of nodes), whereas the complexity of evaluation of (6.2) is much larger, namely, $\Theta(n)$). Finally, let us note that in this approach the embedding e_i that is learned corresponds to $e_{s,i}$ in **Word2Vec**.

Node2Vec is another popular algorithm. In **Node2Vec** a parameterized random walk is considered. Depending on the parameterization, the random walks are respectively biased towards the following two extremes:

- **Breadth-First Search** (**BFS**): walks tend to stay near the initial node, mimicking the **BFS** algorithm, or

- **Depth-First Search** (**DFS**): walks tend to move away from the initial node, mimicking the **DFS** algorithm.

The **DFS** walks tend to preserve the macro-view neighbourhood of a node while **BFS** aims at preserving the micro-view. The user may smoothly move between these two extremes by changing the parameters of **Node2Vec**, as required by a given application. In this algorithm, the approximation of probabilities follows (6.2) with two adaptations. It is assumed that $\mathbf{E}_t = \mathbf{E}_s$. Moreover, in order to solve the problem of $\Theta(n)$ complexity of evaluation of (6.2), a **negative sampling** approximation procedure is applied. In short, the main idea is that when using maximum likelihood estimation to approximate the denominator in (6.2) in each step of the optimization, instead of considering all nodes in the graph, we simply restrict ourselves to all nodes observed in the context window and sample only nodes that are not observed (hence the word *negative* in the name of the procedure).

Deep Learning Algorithms

There are several deep learning methods that have successfully been used to produce node embeddings. These algorithms are quite complex so we will not provide fully detailed explanations. Our goal is merely to provide a brief overview of a few of those algorithms to give the reader a taste of this flavour of graph embeddings.

The first algorithm is an **autoencoder**, a type of artificial neural network that is a commonly used in deep learning to represent complex objects such as images. The goal is to provide a low dimension representation that allows for the original object to be reconstructed as accurately as possible from its low dimensional representation. The aim of an autoencoder is to train the network to ignore signal "noise." The reductionary element is learnt at the same time as the reconstructionary element while the autoencoder tries to generate from the reduced encoding a representation that resembles its original input as closely as possible. Autoencoders are trained to minimize reconstruction errors (such as squared errors), which is often referred to as the loss function.

Structural Deep Network Embedding (SDNE) belongs to this family of algorithms. It aims to preserve both the first and the second order proximity. Recall that the first order proximity is derived directly from weights of the edges while the second order indicates similarity between nodes' neighbourhoods—see Section 6.2 for formal definitions. In the case of **SDNE**, the autoencoder takes the node's adjacency vector as input and tries to learn its embedding. As usual, let k be the dimension of the embedding. Formally, the **encoder** $E \colon \mathbf{R}^n \to \mathbf{R}^k$ is an ℓ-layer neural network that takes n-dimensional input and produces a k-dimensional embedding. On the other hand, the **decoder** is a function $D \colon \mathbf{R}^k \to \mathbf{R}^n$. The two functions, E and D, together form the **autoencoder**. If one denotes the adjacency vector corresponding to node v_i with a_i (if the adjacency matrix $\mathbf{A}$ is symmetric, then a_i is its i-th row or column), then $e_i = E(a_i)$ is the desired embedding. As mentioned above, the goal of the autoencoder is to get $D(E(a_i))$ as close to a_i as possible, which assures that the embedding correctly captures the graph's second-order proximity. It is left to define what exactly is meant by *close* in this context.

Let us define a diagonal matrix $\mathbf{B}_i$, which has 1 in position $b_{j,j}$ if $a_{i,j} = 0$ and $b_{j,j} = \beta$ for some parameter $\beta > 1$ if $a_{i,j} > 0$. The quality of an embedding for node v_i is then measured by the following function $\|(a_i - D(E(a_i)))\mathbf{B}_i\|^2$. As a result, more weight is put on dimensions of the embedding where there is a link between nodes. Now, in order to ensure that the embedding also preserves the first-order proximity, we use the approach of **LEM** and additionally require that $\mathrm{tr}(\mathbf{ELE}^T)$ is small. Therefore, in order for an embedding to simultaneously preserve the first and the second order proximity, we find functions E and D that minimize the following loss function:

$$\sum_{i=1}^{n} \|(a_i - D(E(a_i)))\mathbf{B}_i\|^2 + \alpha \cdot \mathrm{tr}(\mathbf{ELE}^T) + L_{\mathrm{reg}},$$

where parameter α controls the relative importance of the first and the second order proximity measures as determined by the user, and L_{reg} is a penalty function applied to parameters of neural network autoencoder to prevent overfitting.

The next embedding algorithms that we highlight use a similar approach to the **Recursive Feature Extraction (ReFeX)** procedure. This is a fairly general approach and we will concentrate on providing a high-level perspective on the intuitions behind the algorithm. A concrete implementation of this flexible approach will be discussed in Section 6.7 where experiments are presented. Initially, each node in a graph has some set of values assigned to it, which can be some external metadata or some property of the node such as its degree. Let f_i^0 denote these initial features associated with node v_i, and let f^0 be the corresponding vector of such features (so f^0 is a vector of vectors). In each step of the algorithm, we first perform an aggregation using some function a, typically the mean or the sum. Formally, in step $s \in \mathbb{N}$, we compute $f_i^s = a(\{f_j^{s-1} : j \in N(v_i)\})$ to get f^s (again, it is a vector of vectors). Next, we perform a process of feature pruning. In the extreme case, one may use the complete set of feature vectors for each node, that is, the set $\{f^j : j \in [s] \cup \{0\}\}$. However, it is better to select some subset I^s of features that provide information which is not redundant; for example, if two features f^a and f^b are very similar, only one of them is retained using some rule. The algorithm terminates after some predefined number of steps t, or when at step t we have $I^t \subseteq \bigcup_{j=0}^{t-1} I^j$, that is, no new features are retained in step t. After termination, vectors from the set $\bigcup_{j=0}^{t} I^j$ are taken to form rows of a matrix $\mathbf{E}$, for example, by concatenation of vectors contained in them. As a particular application of this general approach, the i-th column of $\mathbf{E}$ can be seen as an embedding of node v_i.

The idea of recursion and aggregation described above is used in deep learning embedding algorithms such as **Graph Convolution Network** (**GCN**) and **GraphSAGE**. Their general procedure is akin to the following approach. We fix the initial embedding e_i^0 for node v_i to be some node attributes provided by the user such as node metadata or some graph statistics. For each node v_i, the algorithm performs the following operation K times. In step $s \in [K]$, the following is computed:

$$e_{N(i)}^s = a_s\Big(\{e_j^{s-1} : j \in N(v_i)\}\Big),$$

where a_s is some **aggregator** function. Next, new knowledge learnt in this step is incorporated by computing $c(e_i^{s-1}, e_{N(i)}^s)$, where c is some **combine** function. The result is fed to layer L_s of a dense neural network that makes a non-linear transformation of the input and, at the same time, constrains the dimensionality of the output. As a result, the embedding of node v_i for step s is $e_i^s = L_s(c(e_i^{s-1}, e_{N(i)}^s))$ and, finally, e_i^s is normalized. The result of the iterative application of this process, e_i^K, is taken as the final embedding of node v_i.

In this general procedure, functions a_s, c and L_s may have various parameters that can be tuned to increase the quality of the final embedding. In an unsupervised setting, they are typically optimized to reconstruct a pairwise similarity matrix $\mathbf{S}$ supplied by the user. However, note that exactly the same model architecture can be used in a supervised context. In this case, for each node v_i, the final embedding e_i^K can be used as a predictor of some feature of node v_i (known in the training sample; for example, some label of the node) and the parameters of the functions a_s, c and L_s are chosen to maximize the quality of this prediction.

The concrete implementations of this general procedure, such as **GCN** and **GraphSAGE**, differ in how functions a_s, c and L_s are defined. For example, in **GCN** function c is a weighted sum and a_s is an element-wise mean. On the other hand, in **GraphSAGE**, function c is a vector concatenation and for a_s the authors of the algorithm consider different options such as element-wise mean, max-pooling neural networks, or LSTM networks.

Note that, in comparison to the embedding algorithms discussed earlier, the deep learning based approaches can have four kinds of advantages, depending on details of their implementations: (1) because of the recursion used, they consider deep rather than shallow structure around each node, (2) they can take into account node attributes passed explicitly by the user, (3) the models can be applied outside the training set, which is important, for example, in the context of evolving networks, (4) the learning process can be supervised so that the embedding is optimized to maximize its predictive power.

Miscellaneous Algorithms

There are over 100 algorithms proposed in the literature for node embeddings. Most of them fall into one or more of the categories defined above but some propose a different approach. One such algorithm is **LINE**, which explicitly defines two functions to encode the first and the second order proximity. In order to capture the first order proximity, the joint probability distribution is defined for a pair of nodes based on their embeddings:

$$p_e^{(1)}(v_i, v_j) = \frac{1}{1 + \exp(-e_i^T e_j)}$$

under the constraint that $\sum_{i=1}^n \sum_{j=1}^n p_e^{(1)}(v_i, v_j) = 1$. This probability distribution is to be compared with a probability distribution based on the adjacency matrix $\mathbf{A}$:

$$p_A(v_i, v_j) = \frac{a(v_i, v_j)}{\sum_{i=1}^n \sum_{j=1}^n a(v_i, v_j)}.$$

Our goal is to find an embedding that minimizes the **Kullback-Leibler divergence** (sometimes called **relative entropy**) between the distributions $p_e^{(1)}$ and p_A that is a standard measure of how one probability distribution is different from a second, reference probability distribution. After simplification, this divergence can be expressed as follows:

$$- \sum_{v_i v_j \in E} a(v_i, v_j) \ln \left(p_e^{(1)}(v_i, v_j) \right).$$

Note that if two nodes have similar embeddings, then they lead to a similar reconstruction of row/column of matrix $\mathbf{A}$.

The method can be adjusted to measure the graph's second order proximity. In this case, each node v_i is assigned source and target embedding vectors,

$e_{s,i}$ and $e_{t,i}$, and the conditional probability distribution is considered for a target of a random edge sampled from the set of edges having one endpoint in v_i:

$$p_e^{(2)}(v_j|v_i) = \frac{\exp(e_{t,j}^T e_{s,i})}{\sum_{\ell=1}^n \exp(e_{t,\ell}^T e_{s,i})}.$$

This time, our goal is to select the source and the target embeddings in a way that minimizes the divergence between all n such distributions computed for a graph, and their counterparts for the two embeddings. There are many ways to specify the analytical formula for this procedure. In a special case, if one considers **Kullback-Leibler divergence** again and puts a weight $\deg(v_i)$ to a distribution associated with node v_i, then it can be shown that the optimized objective function has the form:

$$- \sum_{v_i v_j \in E} a(v_i, v_j) \ln \left(p_e^{(2)}(v_j|v_i)\right).$$

As optimization of the objective function for both the first-order and the second-order approaches of the **LINE** algorithm is computationally challenging, in both cases negative sampling is used in practice. A similar technique was described above for the **Node2Vec** algorithm.

6.4   Unsupervised Benchmarking Framework

In the previous section, we mentioned various embedding algorithms using different tools to preserve various proximity measures between nodes, including those mentioned in Section 6.2. However, this is just the tip of the iceberg; there are many more algorithms and the list constantly grows. Moreover, many of these algorithms have various parameters that can be carefully tuned to generate embeddings in different dimensions. The main question we try to answer in this section is: how do we evaluate these embeddings? Which one is the best and should be used?

In order to answer these questions, we propose a general framework that assigns a "divergence score" to each embedding which, in an unsupervised learning fashion, distinguishes good from bad embeddings. In order to benchmark embeddings, we generalize the **Chung-Lu** random graph model we saw in Section 2.5 to incorporate geometry. Let us start with this model.

Geometric Chung-Lu Model

In the Geometric Chung-Lu model we are not only given the expected degree distribution of a graph G on the set of nodes $V = \{v_1, v_2, \dots, v_n\}$,

$$\mathbf{w} = (w_1, w_2, \dots, w_n) = (\deg_G(v_1), \dots, \deg_G(v_n)),$$

but also an embedding of nodes of G in some k-dimensional space, expressed as a function $\mathcal{E} : V \to \mathbb{R}^k$. In particular, for each pair of nodes, v_i, v_j, we know the distance between them:

$$d_{i,j} = \text{dist}(\mathcal{E}(v_i), \mathcal{E}(v_j)).$$

This framework assumes that good embeddings should allow us to reconstruct edges of graph G in an unbiased way. Therefore, it is desired that the probability that nodes v_i and v_j are adjacent to be a function of $d_{i,j}$, that is, to be proportional to $g(d_{i,j})$ for some function g. The function g should be a decreasing function as long edges should occur less frequently than short ones. There are many natural choices such as $g(d) = d^{-\beta}$ for some $\beta \in [0, \infty)$ or $g(d) = \exp(-\gamma d)$ for some $\gamma \in [0, \infty)$. We use the following, normalized function $g : [0, \infty) \to [0, 1]$: for a fixed $\alpha \in [0, \infty)$, let

$$g(d) := \left(1 - \frac{d - d_{\min}}{d_{\max} - d_{\min}} \right)^\alpha,$$

where

$$d_{\min} = \min\{\text{dist}(\mathcal{E}(v), \mathcal{E}(w)) : v, w \in V, v \neq w\}$$
$$d_{\max} = \max\{\text{dist}(\mathcal{E}(v), \mathcal{E}(w)) : v, w \in V\}$$

are the minimum, and respectively the maximum, distance between nodes in embedding $\mathcal{E}$. Clearly, $g(d_{\min}) = 1$ and $g(d_{\max}) = 0$ (see Figure 6.2); in the computations, we can use clipping to force $g(d_{\min}) < 1$ and/or $g(d_{\max}) > 0$ if required.

Let us make some observations about the parameter of the model, $\alpha \in [0, \infty)$. Note that if $\alpha = 0$ (that is, $g(d) = 1$ for any $d \in [0, \infty)$ with $g(d_{\max}) = 0^0 = 1$), then we recover the original **Chung-Lu** model as the pairwise distances are neglected. Moreover, the larger parameter α, the larger the aversion to long edges.

The **Geometric Chung-Lu** model is the random graph $G(\mathbf{w}, \mathcal{E}, \alpha)$ on the set of nodes $V = \{v_1, \ldots, v_n\}$ in which each pair of nodes v_i, v_j, independently of other pairs, forms an edge with probability $p_{i,j}$, where

$$p_{i,j} = x_i x_j g(d_{i,j})$$

for some carefully tuned weights $x_i \in \mathbb{R}_+$. The weights are selected such that the expected degree of v_i is w_i; that is, for all $i \in [n]$

$$w_i = \sum_{j \in [n], j \neq i} p_{i,j} = x_i \sum_{j \in [n], j \neq i} x_j g(d_{i,j}).$$

Additionally, we set $p_{i,i} = 0$ for $i \in [n]$.

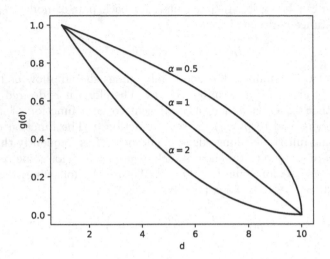

FIGURE 6.2
Function $g(d)$ for $d_{\min} = 1$, $d_{\max} = 10$, and parameter $\alpha \in \{0.5, 1, 2\}$.

It is possible to show that the selection of weights is unique, provided that the maximum degree of G is less than the sum of the degrees of all the other nodes. Since each connected component of G can be embedded independently, we may assume that G is connected and so the minimum degree of G is at least 1. As a result, this very mild condition is trivially satisfied unless G is a star on n nodes.

Framework

Suppose that we are given a graph $G = (V, E)$ on n nodes with degree distribution $\mathbf{w} = (w_1, w_2, \ldots, w_n)$ and an embedding of its nodes to k-dimensional space, $\mathcal{E} : V \to \mathbb{R}^k$.

In a good embedding, one should be able to predict most of the network's edges from the coordinates of the nodes. Formally, it is natural to expect that if two nodes, say u and v, are far away from each other (that is, dist($\mathcal{E}(u), \mathcal{E}(v)$) is relatively large), then the chance that they are adjacent in the graph is less than another pair of nodes that are close to each other. But, of course, in any real-world network there are some sporadic long edges and some nodes that are close to each other yet are not adjacent. In other words, we do not want to pay attention to local properties such as the existence of particular edges (this is the microscopic point of view) but rather evaluate some global properties such as the density of some relatively large subsets of nodes (this is the macroscopic point of view). In other words, one can think of this approach as the goodness-of-fit test for the provided embedding, rather than simply checking its predictive power. Relating it to classical binary classifiers in machine learning, its aim is similar to the well-known **Hosmer-Lemeshow** test for logistic regression. So, the question is: how can we evaluate if the global structure is consistent with our expectations and intuition without considering individual pairs?

The approach we take is as follows. We identify dense parts of the graph by running some stable graph clustering algorithm. The clusters that are found will provide the desired macroscopic point of view of the graph. Note that for this task we only use information about the graph G; in particular, we do not use the embedding $\mathcal{E}$ at all. We then consider the graph G from a different point of view. Using the **Geometric Chung-Lu** model, based on the degree distribution $\mathbf{w}$ and the embedding $\mathcal{E}$, we compute the expected number of edges within each cluster found earlier, as well as the expected number of edges between them. The embedding is scored by computing a divergence score between these expected number of edges and the actual number of edges present in G. This approach falls into a general and commonly used method of *statistical inference*, in our case applied to the **Geometric Chung-Lu** model. With these methods, one fits a generative model of a network to observed network data, and the parameters of the fit tell us about the structure of the network in much the same way that fitting a straight line through a set of data points tells us about their slope.

Of course, not all embeddings proposed in the literature try to capture edges. Some algorithms indeed try to preserve edges whereas others care about some other structural properties; for example, they might try to map together nodes with similar functions. However, since most of the applications we need to deal with require preserving (global) edge densities, the framework favours embeddings that do a good job from that perspective.

Let us now formally describe an algorithm. Our goal is to assign a "divergence score" to the embedding $\mathcal{E} : V \to \mathbb{R}^k$ of G. The lower the score, the better the embedding. This will allow us to compare several embeddings, possibly in different dimensions.

Step 1: Run some stable graph clustering algorithm on G to obtain a partition $\mathbf{C}$ of set V into ℓ communities $C_1, \ldots, C_\ell$. For this purpose, we use **ECG** discussed in Section 5.4 for unweighted graphs, and **Louvain** for weighted graphs, but other algorithms can be used instead.

Step 2: For each $i \in [\ell]$, let c_i be the proportion of edges of G with both endpoints in C_i. Similarly, for each $1 \le i < j \le \ell$, let $c_{i,j}$ be the proportion of edges of G with one endpoint in C_i and the other in C_j. Let

$$\bar{\mathbf{c}} = (c_{1,2}, \ldots, c_{1,\ell}, c_{2,3}, \ldots, c_{2,\ell}, \ldots, c_{\ell-1,\ell}) \quad \text{and}$$

$$\hat{\mathbf{c}} = (c_1, \ldots, c_\ell) \tag{6.3}$$

be two vectors with a total of $\binom{\ell}{2} + \ell = \binom{\ell+1}{2}$ entries which together sum to one. These **graph vectors** characterize the partition $\mathbf{C}$ from the perspective of the graph G.

Step 3: For a given parameter $\alpha \in \mathbb{R}_+$ and the same partition of nodes $\mathbf{C}$, we consider $\mathcal{G}(\mathbf{w}, \mathcal{E}, \alpha)$, the **Geometric Chung-Lu** model. For each $1 \leq i < j \leq \ell$, we compute $b_{i,j}$, the expected proportion of edges of $\mathcal{G}(\mathbf{w}, \mathcal{E}, \alpha)$ with one endpoint in C_i and the other one in C_j. Similarly, for each $i \in [\ell]$, let b_i be the expected proportion of edges within C_i. That gives us another two vectors

$$\bar{\mathbf{b}}_{\mathcal{E}}(\alpha) = (b_{1,2}, \ldots, b_{1,\ell}, b_{2,3}, \ldots, b_{2,\ell}, \ldots, b_{\ell-1,\ell}) \quad \text{and}$$
$$\hat{\mathbf{b}}_{\mathcal{E}}(\alpha) = (b_1, \ldots, b_\ell) \qquad\qquad (6.4)$$

with a total of $\binom{\ell+1}{2}$ entries which together sum to one. These **model vectors** characterize the partition $\mathbf{C}$ from the perspective of the embedding $\mathcal{E}$.

Step 4: In order to measure how well the model $\mathcal{G}(\mathbf{w}, \mathcal{E}, \alpha)$ fits the graph G, compute the distance Δ_α between the concatenated vector consisting of $\bar{\mathbf{c}}$ and $\hat{\mathbf{c}}$, and the one consisting of $\bar{\mathbf{b}}_{\mathcal{E}}(\alpha)$ and $\hat{\mathbf{b}}_{\mathcal{E}}(\alpha)$. Recall that both concatenated vectors have $\binom{\ell+1}{2}$ entries that sum to one. We used the **JensenShannon divergence** which can be viewed as a smoothed version of the **Kullback-Leibler** divergence we mentioned earlier in this chapter.

Alternatively, in order to change the relative importance of internal and external edges, one may independently compute the distances between the two pairs of vectors, that is, between $\bar{\mathbf{c}}$ and $\bar{\mathbf{b}}_{\mathcal{E}}(\alpha)$, and between $\hat{\mathbf{c}}$ and $\hat{\mathbf{b}}_{\mathcal{E}}(\alpha)$. In this case, we let Δ_α to be a weighted average of the two distances.

Step 5: Select $\hat{\alpha} = \text{argmin}_\alpha \Delta_\alpha$, and define the **divergence score** for embedding $\mathcal{E}$ on G as $\Delta_{\mathcal{E}}(G) = \Delta_{\hat{\alpha}}$.

In order to compare several embeddings for the same graph G, we repeat steps 3–5 above and compare the divergence scores (the lower the score, the better). Let us stress that steps 1–2 are done only once, so we use the same partition of the graph into ℓ communities for each embedding.

6.5   Applications

In this section, we highlight a few of the most common applications of graph embedding algorithms. Of course, the list is not intended to be complete and there are many other important potential applications one might want to explore.

In order to illustrate these applications, we immediately show some simple but exemplary analyses. In one of them (visualization) we use the well-known and small Zachary karate club data set that we already used a number of times in this book (see Section 5.2 for more details). For anomaly detection, we use a new network that is especially suitable for experimentation due to the known ground truth. For the remaining experiments, we consider a synthetic graph G generated with the **ABCD** framework (see Section 5.3). The graph consists of $n = 1,000$ nodes partitioned into 12 communities (having sizes ranging between 50 and 150) and $m = 8,327$ edges that exhibit relatively weak community association ($\xi = 0.6$).

We consider an embedding $\mathcal{E}$ of this graph in 48-dimensional space using the **HOPE** algorithm (see Section 6.3) together with the **Personalized PageRank** proximity measure (see Section 6.2). This embedding was chosen as it gives good divergence results when using the framework presented earlier (see Section 6.4), but as detailed in the companion notebook, several other choices give similar results. The choice of a synthetic graph produced by the **ABCD** framework as a reference model has the benefit that we know exactly how the graph was generated (ground truth discussed in Section 5.2) so that it is possible to precisely assess to what extent the goal of a given task was achieved.

Node Classification

Node classification is an example of a semi-supervised learning algorithm where labels are only available for a small fraction of nodes and the goal is to label the remaining set of nodes based on this small initial seed set. This is a situation often observed in complex networks. For example, in social networks labels might indicate a user's interests, beliefs, or demographic characteristics. There could be many reasons for labels not to be available for a large fraction of nodes. For example, coming back to our example of social networks, a user's demographic information might not be available to protect their privacy. Our task is to infer missing labels based on the small set of labelled nodes and the structure of the graph.

Since embedding algorithms can be viewed as the process of extracting features of the nodes from the structure of the graph, one may reduce the problem to a classical machine learning predictive modelling classification problem for the set of vectors. There are many algorithms, such as logistic regression, k-nearest neighbours, decision trees, support vector machine, etc., for any potential scenario that one might be interested in, including binary, multi-class, and multi-label classifications.

In order to show a practical example of the problem of node classification, we performed the following experiment using the synthetic **ABCD** graph G described in the introduction of this section. The community of each node of this graph is its ground-truth community—recall that there are 12 such communities that partition the set of nodes of G. Each node is represented by its 64-dimensional embedding, which can also be seen as its **feature vector** in machine learning terminology. We partition the set of nodes randomly into a training set (with 25% of the nodes) and a test set (with the remaining

75%). We trained a **random forest classifier** (ensemble learning method for classification, regression, and other related tasks that operate by constructing a multitude of decision trees) on the training set, and applied the model to the remaining nodes. The overall accuracy was found to be 90.9% which is quite good in comparison to a random classifier which provides an accuracy of around 9%.

To provide a more detailed picture of multi-class classification, one often summarizes the results using the **confusion matrix** $\mathbf{C} = (c_{i,j})_{i,j \in [n]}$ where $c_{i,j}$ is the number of nodes in community i classified as being in community j. We illustrate this matrix below; in particular, the proportion of weight on the diagonal is the overall accuracy (again, 90.9%).

$$\mathbf{C} = \begin{bmatrix}
78 & 1 & 4 & 0 & 0 & 2 & 3 & 0 & 3 & 0 & 1 & 0 \\
0 & 77 & 0 & 1 & 1 & 0 & 0 & 1 & 0 & 0 & 0 & 0 \\
1 & 0 & 71 & 0 & 1 & 0 & 0 & 0 & 3 & 0 & 0 & 0 \\
1 & 1 & 1 & 64 & 0 & 1 & 0 & 1 & 1 & 0 & 0 & 0 \\
1 & 0 & 2 & 1 & 77 & 0 & 0 & 1 & 0 & 1 & 0 & 0 \\
0 & 0 & 0 & 1 & 1 & 54 & 0 & 1 & 0 & 0 & 0 & 0 \\
1 & 1 & 0 & 0 & 1 & 3 & 53 & 3 & 2 & 0 & 0 & 0 \\
0 & 0 & 0 & 1 & 1 & 0 & 0 & 42 & 0 & 0 & 0 & 0 \\
0 & 0 & 0 & 0 & 1 & 1 & 0 & 1 & 45 & 0 & 0 & 0 \\
1 & 0 & 0 & 0 & 0 & 1 & 0 & 0 & 1 & 44 & 0 & 0 \\
0 & 0 & 0 & 0 & 0 & 0 & 0 & 0 & 0 & 0 & 42 & 0 \\
1 & 1 & 1 & 0 & 0 & 4 & 2 & 2 & 1 & 0 & 0 & 35
\end{bmatrix}$$

Finally, let us point out that in this approach one can easily combine an embedding based on graph structure with additional information available in node metadata to achieve better predictions. Also, as we noted earlier, some methods such as **GraphSAGE** can be directly adjusted to find embeddings that maximize the classification accuracy and take into account the metadata of nodes and their neighbours.

Node Clustering and Community Detection

In Chapter 5, we discussed various techniques and algorithms for detecting communities at length. Node embeddings provide an alternative tool for clustering related nodes. Indeed, since each node can be associated with a real-valued vector embedded in k-dimensional space, one may alternatively ignore the initial graph and apply some generic clustering algorithm to the set of associated vectors. Clustering points seems to be a much easier task and is a well-studied area of research with many scalable algorithms, such as k-**means** or **DB-scan**, that are easily available for use. Finally, one may want to combine the two clustering techniques which presumably should give better results as node embeddings provide some additional information about the functions or roles of particular nodes, something that is not available with graph clustering alone.

Using the same **ABCD** graph G and the embedding $\mathcal{E}$ as before, we compare the results we get from two graph clustering algorithms (namely, **Louvain** and **ECG** discussed in Section 5.4), and two clustering algorithms using the embedding as feature vectors (namely, k-**means** and **DB-scan**). For k-means, we use the correct number of clusters (12) as well as two underestimations (6 and 9) and two overestimations (15 and 24). Results from 30 runs for each algorithm and each number of clusters ($k \in \{6, 9, 12, 15, 24\}$) are given in Figure 6.3. In order to compare the results with the ground-truth community structure, we use **Adjusted Mutual Information (AMI)** which we introduced in Section 5.3. We see that for k-**means** good results are obtained in embedded space, provided that the correct number of clusters (12) is supplied. For **DB-scan** algorithm, the clustering is learned using hyperparameter tuning. Since this algorithm is deterministic, we do not add it to the boxplot and only report that the **AMI** value is equal to 0.66 for all nodes, and 0.87 if outliers are identified by the algorithm and then excluded.

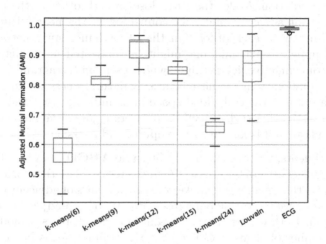

FIGURE 6.3
The comparison of clustering in vector spaces (via embedding $\mathcal{E}$ of graph G and k-**means** algorithm) with graph clustering algorithms.

As a practical note, observe that some algorithms (such as **Louvain** or k-**means**) have a learning phase that is stochastic. It means that the results from a single run of them can be unreliable, as may be observed in Figure 6.3. Therefore, it is standard practice for such algorithms that the best result is carefully selected based on several independent runs of them.

Link Prediction and Missing Links

Node embeddings can also be successfully used to predict missing links or to predict links that are likely to be formed in the future. Indeed, networks are often constructed from the observed interactions between nodes, which may be incomplete or inaccurate. In particular, the situation of missing links is typical in the analysis of biological networks in which verifying the existence of links between nodes requires experiments that are expensive and might not be accurate. Moreover, a task that is closely related to link prediction is the main ingredient of recommendation systems. The goal might be to predict missing friendship links in social networks or to recommend new friends. Another task might be to predict new links between users and possible products that they may like.

Once nodes are embedded in k-dimensional space (assuming that the embedding tries to reflect the probability of edge being present between a pair of nodes; the framework presented in Section 6.4 may be used to make sure it is the case), one may use the distance between the corresponding vectors to make the prediction. Nodes that are close to each other in the embedded space but are not adjacent might get connected in the near future as they seem to be similar to each other. On the other hand, since networks that we typically mine are dynamic, one might be interested in predicting which links will become inactive; for example, which users on Instagram a given user might want to unfollow in the near future. A natural guess would be to pick nodes that are far in the embedded space as it indicates that the nodes are dissimilar. Investigating such behaviour of users is an active area of research of social scientists and is known as homophily and aversion.

For experiments, we continue using the same **ABCD** graph G. This time, we randomly select 10% of the edges of G and remove them, thus forming a new graph G'. From this smaller graph, we consider all pairs of adjacent nodes (the positive class) as well as a random subset of pairs of non-adjacent nodes (the negative class). Both classes have the same number of pairs of nodes (equal to m', the number of edges of G') so that the training set is balanced. Note that it may happen that some pairs that are selected for the negative class are non-adjacent in G' but adjacent in G but in this case it is intentional as in practice we often need to deal with such noisy training sets. We recompute the embedding $\mathcal{E}'$ using graph G' and then use binary operators to combine embeddings of node pairs into feature vectors to be used for model building and prediction. For the classification, we used the **logistic regression model** in which the output is an estimation of the probability for the positive class in the training data set. We trained our model on all data from graph G' (that is, on $2m'$ pairs of nodes coming from the two classes), and applied it to the edges deleted from graph G as well as a random sample of pairs of non-adjacent nodes in G.

We got 56% accuracy on this graph which is just slightly better than random. However, let us point that graph G is very noisy since it was generated with parameter $\xi = 0.6$. As a result, majority of edges are actually noise edges, and so link prediction is an extremely challenging problem for this graph. We re-ran the same experiment with another ABCD graph generated with the same parameters except that $\xi = 0.2$, so most edges are now community edges. The results for this graph are summarized in Figure 6.4 in the form of the **Receiver Operating Characteristic curve (ROC curve)**. In order to draw an **ROC** curve, only the true positive rate (also known as recall) and false positive rate (also known as fall-out) are needed (as functions of some classifier parameter). The best possible prediction method would yield a point $(0,1)$ in the upper left corner, representing 100% sensitivity (no false negatives) and 100% specificity (no false positives)—**perfect classification**. A random guess would give a point along a diagonal line—**line of no-discrimination**. Points above the diagonal represent good classification results (better than random) whereas points below the line represent bad results (worse than random). The **Area Under the ROC Curve (AUC)** provides a measure of separability as it tells us how capable the model is of distinguishing between the two classes. Indeed, **AUC** is bounded from above by 1 and can be interpreted as the probability that a random positive observation has a higher predicted probability than a random negative observation. Also note that since we did class rebalancing for model building, both **ROC** and **AUC** are unaffected by rebalancing of the classes. The accuracy is now 0.76 and the **AUC** is 0.86, much better than random. Moreover, the ROC curve presented in Figure 6.4 is quite steep at the beginning indicating that most of the highest scoring node pairs are truly edges.

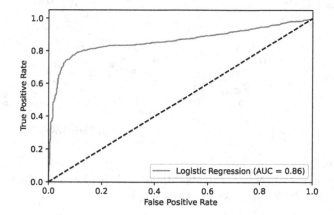

FIGURE 6.4
Results of link prediction on the 1,000 nodes **ABCD** graph with $\xi = 0.2$.

Visualization

Node embeddings can be successfully used for graph visualization. Since nodes can be associated with real-valued vectors in k-dimensional space, one might use any of the various scalable techniques for visualizing high-dimensional points in 2D that are widely available. These generic dimensionality reduction techniques include the well-known **Principal Component Analysis (PCA)** and t-**Distributed Stochastic Neighbour Embedding** (t-**SNE**) techniques, but there are many other approaches available. We personally recommend **Uniform Manifold Approximation and Projection**[1] (**UMAP**), which is a novel manifold learning technique for dimension reduction. It provides a practical scalable algorithm that applies to real world datasets. Finally, as good visualizations require as few long edges as possible, it is recommend to test various embeddings and various parameters and select the one that scores well in the benchmark framework we mentioned in Section 6.4.

In order to allow for clear visualization, we switch now to a much smaller graph, namely, the famous Zachary karate club graph that is supplied in `igraph`. The graph represents external interactions between 34 members of a karate club that, due to a conflict that arose between the administrator and the instructor, eventually split into two groups. In our figures, we represent the two groups via different node shades of grey. In Figure 6.5, we show two different embeddings for this graph (using **UMAP** to project the embedding into two dimensions), respectively, with low and high divergence scores as described in Section 6.4.

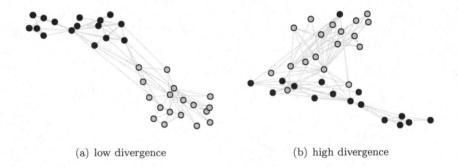

           (a) low divergence   (b) high divergence

FIGURE 6.5
Two-dimensional projections of embeddings with (a) low and (b) high divergence scores for the Zachary karate club graph.

Finally, let us mention that there are some specialized algorithms for visualizing graph data in two (sometimes three) dimensions. Such algorithms are often called *layout* algorithms and can be considered to belong to a special

---

[1]`pypi.org/project/umap-learn/`

family of embedding algorithms. In `igraph`, for example, there are over 15 such algorithms provided. One class of such algorithms are the **force-directed layouts** in which edges can be seen as "springs" applying force to keep adjacent nodes close to each other. Examples of such layout algorithms include the **Kamada-Kawai** and the **Fruchterman-Reingold** algorithms. Another class consists of layouts with some specific pattern such as a grid, circle, sphere, or a tree. There are other possible layout algorithms including multi-dimensional scaling and random layout. We illustrate a few of these algorithms in Figure 6.6 for the same Zachary karate club graph.

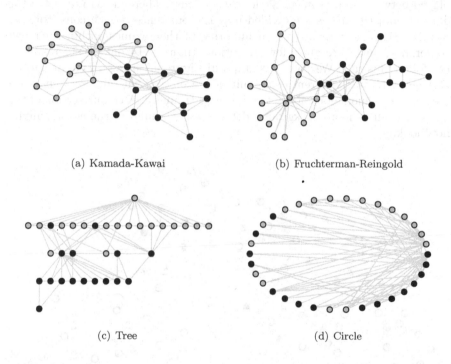

(a) Kamada-Kawai	(b) Fruchterman-Reingold
(c) Tree	(d) Circle

FIGURE 6.6
The Zachary karate club graph presented using various layout algorithms.

Anomaly Detection

For the last example, we illustrate another application of the unsupervised benchmarking framework presented in Section 6.4, the detection of anomalous nodes. The main assumption we make is that a node that is *not* anomalous has most of its neighbours in the same community, or in a small number of communities, while anomalous nodes have neighbours all over the place. In order to detect such nodes, one needs to generate several embeddings and select a good one, $\mathcal{E}$, by applying the benchmarking framework presented earlier in this chapter. Given the embedding $\mathcal{E}$, one may use the corresponding **Geometric Chung-Lu** model to compute the expected fraction of edges between node v and all communities, namely, $P_v = (p_{v,1}, p_{v,2}, \ldots, p_{v,\ell})$ assuming that ℓ communities were found by the framework. From this distribution, we compute the **entropy** $H(P_v) = -\sum_{i=1}^{\ell} p_{v,i} \ln(p_{v,i})$. We repeat this process for all nodes. Given our assumption, high entropy is an indicator of anomalies for the nodes so we can use the entropy scores to rank the nodes from the most likely to the least likely to be anomalous.

We illustrate this approach with a small real graph that depicts games played between American college football teams[2]. There are 115 teams in this dataset, and the teams are divided into 12 conferences which constitute the communities. Most teams play a majority of their games within their own conference, but there are a few exceptions, namely 14 out of 115 teams do not follow this rule. We show this graph in Figure 6.7 using a **Fruchterman-Reingold** layout. Different communities are plotted with different shades of grey, and the anomalous nodes are shown as triangles. A colour version of this figure as well as more background details can be found in the accompanying notebook.

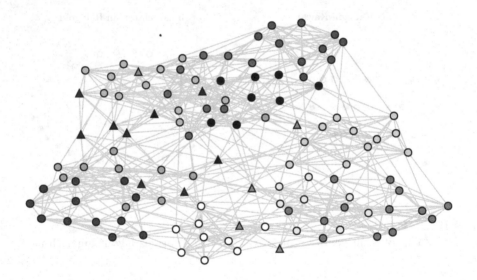

FIGURE 6.7
The American college football graph with communities shown in different shades of grey. Anomalous nodes are shown as triangles.

We ran 60 different variations of the **Node2Vec**[3] embedding algorithm on this dataset, and computed the **Jenssen-Shannon (JS)** divergence scores

[2] www-personal.umich.edu/ mejn/netdata/
[3] snap.stanford.edu/node2vec/

using the unsupervised benchmarking framework. In practice, we would only retain the best embedding(s), but for illustration purpose, we computed the area under the **ROC** curve (**AUC**) when using each of the embeddings to rank the nodes with respect to the entropy score to detect anomalies, where the **ground-truth** is given by the 14 known anomalous nodes. We compare the **AUC** and the **JS** divergence score for those embeddings in Figure 6.8(a), where we clearly see that the best embeddings (with small **JS** divergence score) yield **AUC** values very close to 1 (that is, perfect ranking). Moreover, we also see the quality of this algorithm degrades for less accurate embeddings. In Figure 6.8(b), we plot the distribution of the entropy values for the regular and anomalous nodes when using the best embedding with respect to the benchmarking framework. The difference in entropy values is clearly seen. One variation of this approach is to consider several good embeddings identified via the **JS** divergence score, rank the nodes with respect to entropy for each embedding, and compute the average rank. This will hopefully allow us to investigate different aspects of the data captured by different embeddings. We illustrate this for the football graph in the accompanying notebook.

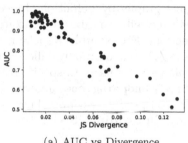
(a) AUC vs Divergence

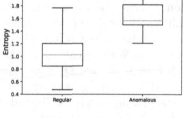
(b) Comparing entropy

FIGURE 6.8
Comparing results from 36 different embeddings with **Node2Vec**. In (a) we show the relation between the **JS** divergence score and the resulting **AUC** when ranking the nodes as anomalous. In (b), we compare the entropy values for regular vs. anomalous nodes when using one of the best embedding.

6.6 Other Directions

It should be clear by now that embedding graph is a complex task and here we only scratch the surface. In this section, we briefly mention other directions that are important from a practical point of view.

Directed Graphs

The main difference when considering directions of edges is that while distance between nodes in the embedded space is symmetric, the probability of an edge may vary depending on the choice of the source and the target node. One of the key applications of embeddings of directed graph is *link prediction*. As a result, several algorithms embed nodes with a concatenation of two vectors that reflect the source and, respectively, the target roles of that node, allowing for asymmetric edge probabilities between the two corresponding nodes. **HOPE**, which we have already used for undirected graphs, is an example of such an algorithm that can handle directed graphs. Another family of algorithms aims at learning two objects: the spatial embedding of the nodes, but also a vector field in embedded space that indicates the dominant direction of edges depending on the spatial position.

Signed Graphs

A **signed graph** is a graph $G = (V, E)$ in which each edge has a positive or negative sign, that is, the set of edges E is partitioned into two sets, E_+ and E_-. Many complex networks are signed graphs by definition. Indeed, relations between users on social media sites often reflect a mixture of positive (friendly) and negative (antagonistic) interactions. For example, a social news website Slashdot.org (often abbreviated as /.) that originally billed itself as "News for Nerds. Stuff that Matters.", allows their users to specify other users as friends or foes. Epinions.com (currently Shopping.com) allows users to mark their trust or distrust to other users on product reviews. There are some embeddings, such as **Signed Network Embedding** (**SNE**), that are crafted to deal with such networks.

Embedding Edges

In contrast to node embedding, edge embedding aims to represent an edge as a low-dimensional vector. However, it is not obvious how to naturally define edge-level similarity as an edge consists of a pair of nodes. In particular, an edge can be directed and, if this is the case, then the direction should be taken into account. Moreover, a good embedding algorithm should incorporate the importance of edges when embedding them into $\mathbb{R}^k$. Embedding nodes is a relatively well-established field now but embedding edges requires more work. Having said that, there are two natural approaches that can be used.

For tasks such as link prediction, where a classifier needs to be trained on both positive (existing) and negative (not existing) edge representations, in order to embed a pair of nodes uv (again, representing edge or non-edge), one may simply use some aggregation function such as the average of $\mathcal{E}(u)$ and $\mathcal{E}(v)$. However, since node embedding algorithms inherently focus on nodes, using such aggregations may not generate good results. Alternatively, one may consider the **line graph** of an undirected graph $G = (V, E)$ that is another graph $L(G)$ that represents the adjacencies between edges of G. Formally, the set of nodes of $L(G)$ is E, and two nodes e_1, e_2 are adjacent in $L(G)$ if the corresponding edges are incident in G. In order to embed edges of G one may simply use the embedding of nodes in $L(G)$.

Multi-Layered Graphs

In some applications, we are provided with ℓ graphs $G_i = (V_i, E_i)$, $i \in [\ell]$, with overlapping sets of nodes. In particular, when $V_1 = V_2 = \ldots = V_\ell$, we may view it as one graph on the set of nodes V consisting of multiple "layers." For example, in protein-protein interaction networks derived from different tissues (say, from brain and liver tissues), some proteins occur across multiple tissues.

A good embedding algorithm should embed one of the layers taking into account information coming from other layers. For example, **OhmNet** algorithm introduces a penalty term that tries to tie the embeddings across layers. The loss function can be augmented as follows:

$$\mathcal{L}' = \mathcal{L} + \lambda \sum_{v \in V_1 \cap V_2} \|\mathcal{E}_{G_1}(v) - \mathcal{E}_{G_2}(v)\|,$$

where $\mathcal{L}$ is the original loss function, λ is the regularization strength, and $\mathcal{E}_{G_i}(v)$ is the embedding of v in layer i.

Moreover, quite often there is some natural hierarchy between layers. For example, in protein-protein interaction graphs derived from various tissues, some layers correspond to interactions throughout large regions whereas other ones are more detailed and fine-grained. In such situations, embeddings can be obtained by recursively applying the regularization equation following the hierarchy between layers.

Hyperbolic Spaces

Graph Convolutional Neural networks (**GCNs**), which we discussed earlier, embed nodes in Euclidean space. Such embedding algorithms applied to some real-world graphs with scale-free or hierarchical structure produced outcomes that incur a relatively large distortion, that is, the graph distances between pairs of nodes could not be accurately estimated based on the Euclidean distances between the corresponding pairs of embeddings. Initial experiments show that embeddings in hyperbolic geometries produce smaller distortion and so they offer a possible alternative for such families of graphs. The **Hyperbolic Graph Convolutional Neural Network** (**HGCN**) is the first inductive hyperbolic counterpart of **GCN** that leverages both the expressiveness of **GCNs** and hyperbolic geometry to learn inductive node representations for hierarchical and scale-free graphs.

Embedding Graphs

Graph embedding is a technique that aims to map the entire graph to a point in a vector space. Graph embedding methods can be vaguely divided into two different categories: explicit graph embeddings and implicit graph embeddings, which are also known as graph kernels. As the name indicates, explicit graph embeddings provide algorithms that return embeddings $\mathcal{E} : \mathcal{F} \to \mathbb{R}^k$, where $\mathcal{F} = \{G_i = (V_i, E_i) : i \in [n]\}$ is a family of graphs we wish to embed. Such algorithms might be randomized or deterministic but the resulting embedding is immediately usable. On the other hand, implicit graph embeddings only provide a pairwise similarity measure between graphs. This might be enough for the application at hand, but in case it is not one can try to use pairwise distances to find an embedding that preserves them as much as possible.

The first family includes graph probing, which measures the frequency of specific substructures in graphs which tries to capture both the content and topology of the network. In particular, we might want to count the number of **graphlets**, that is, small connected non-isomorphic induced subgraphs present in a large network. Another approach is based on spectral graph theory and aims to analyze the structural properties of graphs in terms of the eigenvectors/eigenvalues of their adjacency or Laplacian matrices. The third class of methods is inspired by dissimilarity measurements. For example, we might want to estimate the distance from a given graph to a number of carefully pre-selected prototype graphs. Finally, motivated by the recent advancements in deep learning and neural networks, one may want to utilize neural networks to obtain a representation of graphs as vectors. One such approach is the **Graph2Vec** algorithm which we will use in Chapter 9.

The second family uses **graph kernels** to evaluate the similarity between a pair of graphs G and G' by recursively decomposing them into atomic substructures (for example, random walks, shortest paths, graphlets, etc.). Then, one may define a similarity kernel function over the selected substructures (for example, counting the number of common substructures across G and G') and use it as a similarity measure. Subsequently, a typical approach is to, for example, apply some kernel method such as **Support Vector Machines** (**SVMs**) to perform classification or clustering, depending on the application at hand.

6.7   Experiments

Embedding and Supervised Learning

In Section 6.5, we illustrated several applications of node embeddings for which we selected the embedding to be used based on the unsupervised framework presented in Section 6.4. While the framework generally selects good embeddings for a variety of applications, in the case of supervised learning, we can often do better by carefully selecting the embedding based on the true objective function for the task at hand. This is typically done by dividing the labelled data into training, validation, and test sets. We use the training and the validation set to select the best model and parameters; in this case, the best embedding. The error rate is then obtained by applying the selected model to the test set. We illustrate this process by re-visiting the node classification problem for a larger instance of the **ABCD** graph.

We begin by partitioning the set of nodes into three bins, with 25% of nodes in the training set, 25% in the validation set, and 50% in the test set. We trained 70 different embeddings: we used **HOPE** with the 4 different proximity measures presented in Section 6.2 (Katz Index, Personalized PageRank, Common Neighbours, Adamic-Adar), **LEM**, and **Node2Vec** with 5 different combinations of the parameters p and q (corresponding to bias toward BFS or, respectively, DFS walks). We also considered 7 choices of dimensions ranging from 2 to 48 in each case. For each embedding, we trained the random forest classifier using the training data and computed the accuracy on the validation data.

In Figure 6.9, we show the results for the top-10 models obtained via supervised learning using the accuracy score on the validation set. We also present results for the top-10 models based only on the (unsupervised) divergence score from the framework. For both scenarios, the final accuracy score is computed for the test set. Based on the figure, we conclude that we obtain the best results when we use the true objective function and are given labelled data (here, the accuracy), while the results using the unsupervised framework are also very good. Indeed, the mean accuracy for all cases considered is about 61%, and random classification gave us 9% accuracy, which is worse than the results we got with all of the 70 embeddings. Another way to see this is presented in Figure 6.10, where we compare the accuracy score on the test set with the accuracy on the validation set (supervised) and, respectively, the divergence score (unsupervised). While they both show negative correlation, it is clearer in the supervised context.

Role Embedding

In the previous experiment, we were essentially interested in embeddings that preserved the proximity of adjacent or nearby nodes, thus preserving communities. For the next example, we consider a different type of objective where we try to group nodes that have similar roles within the graph.

ReFeX is a framework designed to recursively learn node features in a graph, which can be followed by dimension reduction yielding a vector space representation for each node, as in node embedding. This fairly general approach was already discussed in Section 6.3 where a high-level intuition behind the algorithm was provided. Here, we illustrate this algorithm using the

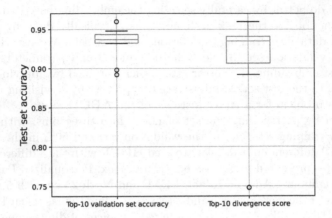

FIGURE 6.9
Results with top-10 models for supervised learning on the **ABCD** graph. For
the left plot, we use the (supervised) accuracy on the validation set to select
the best models. For the right plot, we used the (unsupervised) divergence
score from the framework.

specific implementation from the Python `graphrole`[4] package. The following
features are computed for each node: the degree, the number of edges inside
its ego-net, and the number of edges going out of its ego-net. For the recursion, the following aggregation functions are used: for the degree we consider
both the sum and the mean; for the other features we use the mean. All the
features are concatenated for each node. This is followed by a dimension reduction method known as **Role eXtraction** (**RolX**) that uses non-negative
matrix factorization in order to map each node to a structural representation.
In a nutshell, given the user-supplied parameter k (the number of roles), each
node is mapped to a probability distribution over those roles.

We illustrate this process for the Zachary karate club graph. The **ReFeX**
process, as described above, yielded 7-dimensional feature vectors for each
node. We then mapped the resulting features for each node into 3-dimensional
space corresponding to $k = 3$ "roles." The result is shown in Figure 6.11. We
colour each node with respect to its role, that is, the highest probability role.
From the figure, we see that the roles can roughly be categorized as hub nodes,
periphery nodes, and roles which are in-between.

---

[4]pypi.org/project/graphrole/

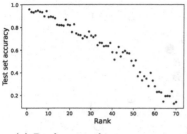

(a) Rank w.r.t. the accuracy score

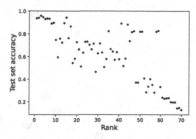

(b) Rank w.r.t. the divergence score

FIGURE 6.10
Comparing test set accuracy with (a) the accuracy (supervised) score on the validation set, and (b) the divergence (unsupervised) score for the **ABCD** graph.

6.8 Practitioner's Corner

The most common usage of graph embedding is to learn a mapping from the set of nodes V into some k-dimensional vector space, where typically $k \ll |V|$. This can be viewed as the process of learning feature vectors for each node. It is motivated by the fact that mining data in vector spaces is often easier to perform due to the availability of tools and techniques. It is also easier to obtain a representative sample of the data in vector spaces, something that is non-trivial to do for graphs.

We saw that most node embedding algorithms roughly fall into three categories: those based on linear algebra (mainly applied to the adjacency matrix), those based on random walks, and more recent methods based on deep learning. As there are many methods proposed in the literature, the user is usually faced with the challenge of selecting the most appropriate one for the problem at hand.

As with most machine learning techniques and tools, problems in graph mining can be grouped into two categories: unsupervised and supervised, and the recommended procedure for selecting an appropriate embedding depends on which category the problem belongs to. The first category refers to *unsupervised* learning, which includes applications such as visualization, clustering, and community detection. For visualization, mainly in two dimensions, there exist several good specialized *layout* algorithms but one may also use an embedding algorithm and then reduce the dimension to 2 or 3 using **UMAP** (see Section 6.5). It is generally difficult to assess the performance of unsupervised algorithms due to a lack of "ground-truth" data. This includes visualization that can be evaluated by mere eyeballing but, of course, this cannot be done for hundreds of embeddings. To that effect, the framework based on the **Geometric Chung-Lu** model can be efficiently used (see Section 6.4). The second

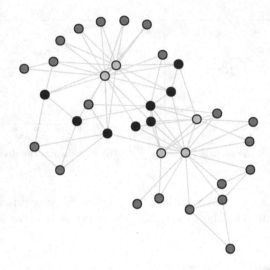

FIGURE 6.11
Role embedding via **ReFeX** and **RolX** for the Zachary karate club graph.

category of problems are *supervised* in the sense that at least some notion of ground-truth information is available; this includes problems such as link prediction and node classification. For such problems, the choice of embedding should be guided by an objective function tuned for the problem at hand and the ground-truth data.

Finally, let us mention that graph embedding is a very recent and active research area; this chapter mainly covered node embedding techniques but there are several other related topics such as edge embedding and embedding the whole graphs.

6.9   Problems

In this section, we present a collection of potential practical problems for the reader to attempt.

1. Generate the **ABCD** graph on $n = 1,000$ nodes with $\gamma = 2.5$, $\beta = 1.5$, and $\xi = 0.2$. Pick 2 of your favourite embedding algorithms, with at least one of them being non-deterministic. For each algorithm and each dimension $(4, 8, 16, 32, 64, 128)$:

   a. independently generate $k = 30$ embeddings,

    b. evaluate the quality of each embedding by computing the divergence score (see the unsupervised benchmarking framework),

    c. compute the mean score (over 30 embeddings) and standard deviation.

Which dimension produced the best embeddings (according to the framework) and which one is the most stable?

2. For the best and the worst embedding (based on the unsupervised benchmarking framework evaluating all $2 \cdot 6 \cdot 30 = 360$ embeddings that were generated in the previous problem), run the k-means algorithm using the known number of communities from the **ABCD** graph (ground truth). Compare the quality of the embeddings using **AMI**, **ARI**, and **AGRI**.

3. Take a ring of cliques (Figure 5.12). For each algorithm you picked in the previous problem, and each dimension $(4, 8, 16, 32, 64, 128)$:

    a. independently generate $k = 30$ embeddings,

    b. evaluate the quality of each embedding by computing the divergence score (see the unsupervised benchmarking framework).

Select the best embedding and the worst embedding (according to the framework) from all embeddings you created ($2 \cdot 6 \cdot 30 = 360$ embeddings), reduce the dimension to 2 using **UMAP**, and plot both of them highlighting the cliques with colours.

4. Repeat the above experiment (plotting the best and the worst embedding) for the giant component of the subset of the European Grid network used to generate Figure 1.2. Additionally,

    a. find the communities of that graph using **ECG** and colour nodes accordingly,

    b. for each pair of nodes, compute the distance in the embedded space and the geographical distance and create a scatter plot,

    c. partition all pairs of nodes into 10 buckets of equal size based on their distance (that is, the first bucket consists of pairs of nodes that are close to each other and the last one pairs of nodes that are far from each other); compute the number of edges that fall into each bucket. What can you conclude?

5. Consider the undirected, unweighted version of the Airport graph that was considered in the previous chapter. Run a ReFeX/RolX analysis as we did for the Zachary karate club graph, using `max_generations=3` and splitting the nodes into 3 roles. Compare the distributions of degrees and the betweenness centrality amongst the 3 groups. What can you conclude?

6.10   Recommended Supplementary Reading

- L. Katz. "A new status index derived from sociometric analysis", *Psychometrika* 18(1):39–43, 1953. (Katz Index)

- P. Berkhin, "A survey on PageRank computing", *Internet mathematics* 2.1 (2005), 73–120. (Personalized PageRank)

- L.A. Adamic, E. Adar. "Friends and neighbors on the web", *Social networks* 25.3 (2003), 211–230. (Adamic-Adar)

- S.T. Roweis, L.K. Saul. "Nonlinear dimensionality reduction by locally linear embedding", *Science* 290(5500):2323–2326, 2000. (LLE, Local Linear Embedding)

- M. Belkin, P. Niyogi. "Laplacian eigenmaps and spectral techniques for embed- ding and clustering", In *Nips*, volume 14, pages 585–591, 2001. (LEM, Laplacian Eigenmaps)

- M. Ou, P. Cui, J. Pei, Z. Zhang, W. Zhu. "Asymmetric Transitivity Preserving Graph Embedding", in *KDD* (2016). (HOPE)

- T. Mikolov. "Distributed representations of words and phrases and their compositionality", *Advances in Neural Information Processing Systems* (2013). (Word2Vec)

- A. Grover, J. Leskovec. "Scalable Feature Learning for Networks", in *ACM SIGKDD International Conference on Knowledge Discovery and Data Mining (KDD)*, 2016. (Node2Vec)

- B. Perozzi, R. Al-Rfou, S. Skiena. "DeepWalk: Online learning of social representations", in *KDD*, 2014. (Deep Walk)

- D. Wang, P. Cui, W. Zhu. "Structural deep network embedding", in *Proc. ACM SIGKDD*, 2016, 1225–1234. (SDNE)

- K. Henderson, B. Gallagher, L. Li, L. Akoglu, T. Eliassi-Rad, H. Tong, C. Faloutsos, "It's who you know: graph mining using recursive structural features", in *Proceedings of the 17th ACM SIGKDD international conference on Knowledge discovery and data mining*, August 2011 pp. 663671. (ReFeX, Recursive Feature eXtraction)

- K. Henderson, B. Gallagher, T. Eliassi-Rad, H. Tong, S. Basu, L. Akoglu, D. Koutra, C. Faloutsos, L. Li, "RolX: structural role extraction and mining in large graphs", in *Proceedings of the 18th ACM SIGKDD international conference on Knowledge discovery and data mining*, August 2012, pp. 12311239 (RolX, Role eXtraction)

- T.N. Kipf, M. Welling. "Semi-supervised classification with graph convolutional networks", in *Proc. of ICLR*, 2017. (GCN)

- W.L. Hamilton, Z. Ying, J. Leskovec. "Inductive representation learning on large graphs", in *Advances in Neural Information Processing Systems 30: Annual Conference on Neural Information Processing Systems 2017*, 2017, 1024–1034. (GraphSAGE)

- J. Tang, M. Qu, M. Wang, M. Zhang, J. Yan, Q. Mei. "Line: Large-scale information network embedding", *Proceedings 24th International Conference on World Wide Web*, 2015, 10671077. (LINE)

- B. Kamiński, P. Prałat, F. Théberge. "An Unsupervised Framework for Comparing Graph Embeddings", *Journal of Complex Networks* 8(5) (2020), cnz043. (Unsupervised Benchmarking Framework)

- B. Kamiński, P. Prałat, F. Théberge. "A Scalable Unsupervised Framework for Comparing Graph Embeddings", *Proceedings of the 17th Workshop on Algorithms and Models for the Web Graph (WAW 2020)*, Lecture Notes in Computer Science 12091, Springer, 2020, 52–67. (Unsupervised Benchmarking Framework)

- W. Suhang, T. Jiliang, A. Charu, C. Yi, L. Huan. "Signed Network Embedding in Social Media", *SDM* 2017: 327–335. (SiNE, Signed Network Embedding)

- M. Zitnik, J. Leskovec. "Predicting multicellular function through multilayer tissue networks", *Bioinformatics* 33(14), (2017) 190–198. (OhmNet)

- I. Chami, Z. Ying, C. Ré, J. Leskovec. "Hyperbolic graph convolutional neural networks", in *Advances in Neural Information Processing Systems*, volume 32, 2019. (HGCN, Hyperbolic Graph Convolutional Neural Network)

- A. Narayanan, M. Chandramohan, R. Venkatesan, L. Chen, Y. Liu, "graph2vec: Learning distributed representations of graphs", MLG 2017, *13th International Workshop on Mining and Learning with Graphs*. (Graph2Vec)

- L. McInnes, J. Healy. "UMAP: uniform manifold approximation and projection for dimension reduction", 2018, preprint available at https://arxiv.org/abs/1802.03426. (UMAP)

Graph embedding surveys:

- H. Cai, V. Zheng, K. Chang, "A Comprehensive Survey of Graph Embedding: Problems, Techniques, and Applications", *IEEE Trans. on Knowledge & Data Eng.*, vol. 30, no. 09, pp. 1616-1637, 2018.

- P. Goyal, E. Ferrara, "Graph embedding techniques, applications, and performance: A survey", *Knowledge-Based Systems*, Vol. 151, pp. 78-94 (2018).

- W.L. Hamilton, R. Ying, J. Leskovec, "Representation Learning on Graphs: Methods and Applications", *IEEE Data Eng. Bull.* 40(3): 52-74 (2017).

- D. Zhang, J. Yin, X. Zhu C. Zhang, "Network Representation Learning: A Survey", *IEEE Trans. on Big Data*, vol. 6, no. 01, pp. 3-28 (2020).

The college football graph was first introduced and analyzed in:

- M. Girvan, M.E.J. Newman, "Community structure in social and biological networks", *PNAS* 99(12), (2002) 7821–7826.

7

Hypergraphs

7.1 Introduction

A hypergraph is a natural generalization of a graph in which hyperedges (counterparts of edges in graphs) may consist of any number of nodes, not just two as in the case of graphs. Many networks that are currently modelled as graphs would be more accurately modelled as hypergraphs. This includes the collaboration network in which nodes correspond to researchers and hyperedges correspond to papers that consist of nodes associated with researchers that co-author a given paper. Another example that we used a few times in this book is the GitHub hypergraph in which nodes correspond to users of the platform and hyperedges correspond to repositories linking users that committed to them. Up to this point, we thought about this object as a graph but, arguably, it is more natural to think of it as a hypergraph. In particular, note that in the examples we have given above, a hypergraph representation provides a lossless representation of data while using a graph entails the loss of some important information. For example, if we see three researchers in the collaboration network forming a triangle in a graph, then it is impossible to tell if they jointly co-authored one paper or published three independent papers co-authored by the three pairs of researchers.

Despite the fact that a myriad of problems can be naturally described in terms of hypergraphs and the fact that they were formally defined in the 1960s (and various realizations were studied long before that), the algorithms and tools that work with hypergraphs are still patchy and often not sufficient to deal with a given practical problem at hand. As a result, researchers and practitioners typically create the 2-section graph of a hypergraph of interest (that is, replace each hyperedge of size k with the complete graph on k nodes; formal definition will be introduced soon). After moving to the 2-section graph, one clearly loses some important information about hyperedges of size greater than two, as explained above, and so a belief commonly persists that one can do better by using the knowledge of the original hypergraph. This is an active area of research in mining complex networks and there are no standard tools yet. In order to highlight some recent developments, we decided to focus on aspects related to community detection that the authors of this book are personally involved in.

This chapter is structured as follows. We first introduce some basic definitions related to hypergraphs (Section 7.2). Then, we define a few random hypergraph models that we will need later on (Section 7.3). In order to highlight some recent attempts to mine hypergraphs, we concentrate on detecting communities in such structures (Section 7.4). As usual, we finish the chapter with experiments (Section 7.5) and provide some tips for practitioners (Section 7.6).

7.2   Basic Definitions

A **hypergraph** $H = (V, E)$ consists of V, the **set of nodes**, and E, the **set of hyperedges**. Each hyperedge $e \in E$ is simply a subset of V, that is, $e \subseteq V$. Hence, each graph $G = (V, E)$ is a hypergraph in which each edge has size two, that is, $|e| = 2$ for all $e \in E$. Similarly as for graphs, we will usually assume that the hypergraphs we work with are **simple**, meaning that no nodes are repeated within one hyperedge and there are no parallel hyperedges. Having said that, it will be convenient to deal with multi-sets when generalizing the **Chung-Lu** model to hypergraphs. Moreover, let us mention that some authors call a hyperedge $e \in E$ with $|e| = 1$ a **loop** but it seems more natural (and consistent with the counterpart for graphs) to call a loop an edge with one node repeated twice. On the other hand, if a given application requires parallel hyperedges, one can usually deal with that by assigning weights to hyperedges. A hypergraph is called k-**uniform** if each hyperedge contains precisely k nodes. So a 2-uniform hypergraph is simply a graph, a 3-uniform hypergraph consists of a collection of unordered triples, and so on.

A node $v \in V$ is called **isolated** if it does not belong to any hyperedge, that is, $v \in V \setminus \bigcup_{e \in E} e$. Two nodes in a hypergraph are **adjacent** if there is a hyperedge which contains both of them. Two hyperedges are **incident** if their intersection is not empty. The **star** $H(v)$ centered in node v is the family of hyperedges containing v, that is, $H(v) = \{e \in E : v \cap e \neq \emptyset\}$; $\deg(v) = |H(v)|$ is the **degree** of v. If each node has the same degree, we say that the hypergraph is **regular**, or d-**regular** if for every $v \in V$, $\deg(v) = d$. As for graphs, the **maximum** degree of a hypergraph H is denoted by $\Delta = \Delta(H) = \max_{v \in V} \deg(v)$, and the **minimum** degree of a hypergraph H is denoted by $\delta = \delta(H) = \min_{v \in V} \deg(v)$. The **rank** $r(H)$ of H is the maximum cardinality of a hyperedge in the hypergraph, that is, $r(H) = \max_{e \in E} |e|$; the minimum cardinality of a hyperedge is the **co-rank** $cr(H)$ of H, that is, $cr(H) = \min_{e \in E} |e|$. Note that if $r(H) = cr(H) = k$, then hypergraph H is k-uniform.

Let us fix a pair of nodes, $u, v \in V$. A sequence of nodes $P = (u = w_0, w_1, \ldots, w_\ell = v)$ is called a **path** from u to v (of **length** ℓ) if nodes

w_{i-1} and w_i are adjacent for all $i \in [\ell]$. Similarly to graphs, a **connected component** of a hypergraph H is a maximal subgraph in which any two nodes are connected to each other by a path. If H has precisely one connected component, then we say that H is **connected**; otherwise, we say that H is **disconnected**.

It will sometimes be convenient to represent hypergraphs as graphs despite the fact that the first two graph representations of hypergraphs defined below do *not* preserve all of the information about the hypergraph, that is, one may *not* be able to reconstruct hypergraph H from its projection to these graphs. Let $H = (V, E)$ be any hypergraph.

The **line graph** of H is a graph $L(H) = (V', E')$ such that $V' = E$ and $e_1, e_2 \in V'$ are adjacent if and only if $e_i \cap e_j \neq \emptyset$. Despite the fact that we lose a lot of details about the original hypergraph, some of its properties can be tested on the corresponding line graph. For instance, it is easy to show that the hypergraph H is connected if and only if $L(H)$ is.

Another useful graph is the **2-section** of H which is the graph denoted by $[H]_2$ in which nodes are the nodes of H and two distinct nodes $u, v \in V$ form an edge if they belong to the same hyperedge in H. In fact, $[H]_2$ is often defined as a multi-graph (or, alternatively, as a weighted graph) in which there are $|\{e \in E : \{u, v\} \subseteq e\}|$ parallel edges between u and v, representing the number of hyperedges that contain both nodes. In other words, each hyperedge of size k is replaced with the complete graph on k nodes which may create parallel edges.

The last reduction to graphs we want to mention is *not* lossy, that is, one may reconstruct the original hypergraph from the corresponding graph. The **incidence graph** of a hypergraph H is a bipartite graph $IG(H) = (V', E')$ with the set of nodes $V' = V \cup E$ (V and E are the two bipartite sets), and where $v \in V$ and $e \in E$ are adjacent if and only if $v \in e$. By counting edges in this bipartite graph that are incident to nodes in V and, respectively, that are incident to nodes in E, we get the following observation that holds for any hypergraph $H = (V, E)$:

$$\sum_{v \in V} \deg(v) = \sum_{e \in E} |e|.$$

Finally, let us briefly mention some algebraic definitions that apply to hypergraphs. Let $H = (V, E)$ be a hypergraph with nodes and edges labelled as follows: $V = \{v_1, v_2, \ldots, v_n\}$ and $E = \{e_1, e_2, \ldots, e_m\}$. Suppose that H has no isolated nodes. Then, H has an $n \times m$ **incidence matrix** $\mathbf{B} = (b_{ij})_{i \in [n], j \in [m]}$ where $b_{ij} = 1$ if $v_i \in e_j$; otherwise, $b_{ij} = 0$. The **adjacency matrix** $\mathbf{A} = (a_{ij})_{i,j \in [n]}$ is a $n \times n$ matrix defined as follows: a_{ij} is equal to the number of hyperedges that contain both v_i and v_j, that is, $a_{ij} = |\{e \in E : \{v_i, v_j\} \subseteq e\}|$. Note that the adjacency matrix of a hypergraph is equal to the adjacency matrix of the corresponding **2-section** graph. However, as mentioned above, reduction to the **2-section** graph entails the loss of some properties and the original hypergraph cannot be reconstructed by looking at its adjacency matrix.

7.3   Random Hypergraph Models

As mentioned in Section 2.1, there are at least four reasons why one might want to consider random models. So far, random hypergraphs have received significantly less attention than random graphs. Moreover, they are almost exclusively considered from a theoretical point of view (reason (i)), and researchers mainly concentrate on asymptotic results obtained for random k-uniform hypergraphs, where $k \geq 3$ is a fixed constant. We define a slightly more general model below but it will be straightforward to restrict ourselves to k-uniform hypergraphs.

Let $\mathcal{P} = (p_1, p_2, \ldots, p_k)$ be a sequence of real numbers from $[0, 1]$. The **binomial random hypergraph** $\mathcal{H}(n, \mathcal{P})$ can be generated by starting with the empty hypergraph on the set of nodes $[n] = \{1, 2, \ldots, n\}$. For each $i \in [k]$ and each i-set of nodes $e \subseteq [n]$, $|e| = i$, we independently introduce a hyperedge e in H with probability p_i.

As for graphs, we note that $p_i = p_i(n)$ may (and usually does) tend to zero as n tends to infinity. For example, it is often assumed that $p_i = c_i / \binom{n-1}{i-1}$ for some non-negative constant c_i so that for any $v \in [n]$

$$\mathbb{E} \deg(v) = \sum_{i=1}^{k} \binom{n-1}{i-1} \cdot p_i = \sum_{i=1}^{k} c_i. \tag{7.1}$$

Let us also note that we formally defined the probability distribution over a family of labelled hypergraphs on n nodes. Finally, as mentioned earlier, the **random k-uniform hypergraph** $\mathcal{H}(n, p, k)$ is simply $\mathcal{H}(n, \mathcal{P})$ with $\mathcal{P} = (p_1, p_2, \ldots, p_k)$ such that $p_k = p$ and $p_i = 0$ for $i \in [k-1]$. On the other hand, the **binomial random hypergraph** $\mathcal{H}(n, \mathcal{P})$ can be viewed as a union of k independent **random k-uniform hypergraphs** $\mathcal{H}(n, p_i, i)$, $i \in [k]$.

Many of the properties of $\mathcal{G}(n, p)$ have been generalized without too much difficulty to $\mathcal{H}(n, p, k)$ but, of course, there are some properties that required more advanced ideas (there are also a few famous examples where the corresponding results for hypergraphs were actually easier to prove). Let us mention some results that can be easily generalized from $\mathcal{G}(n, p)$. The following property holds for a fixed $k \geq 3$ and $p = c / \binom{n-1}{k-1}$ for some constant $c \in \mathbb{R}_+$. If $c < 1/(k-1)$, then a.a.s. the largest component is of size $O(\ln n)$. On the other hand, if $c > 1/(k-1)$, then a.a.s. there is a unique giant component of size $\Theta(n)$ and all other components are of size $O(\ln n)$. In fact, the size of the giant component can be obtained as follows. For $c > 1/(k-1)$, let $x = x(c)$ be the solution in $(0, 1/(k-1))$ to

$$xe^{-(k-1)x} = ce^{-(k-1)c}. \tag{7.2}$$

Then, a.a.s. the giant component has size asymptotic to $(1 - (x/c)^{1/(k-1)})n$.

Let us also mention that the connectivity threshold for $\mathcal{H}(n,p,k)$ coincides with the threshold for the minimum degree being at least one, and so the behaviour is the same as for $\mathcal{G}(n,p)$. Let

$$p = \frac{\ln n + c}{\binom{n-1}{k-1}}.$$

Then,

$$\mathbb{P}\Big(\mathcal{H}(n,p,k) \text{ is connected}\Big) \sim \begin{cases} 0 & \text{if } c \to -\infty \\ e^{-e^{-c}} & \text{if } c \in \mathbb{R} \\ 1 & \text{if } c \to \infty. \end{cases}$$

In particular, we get the following corollary of this much stronger result. If $p \cdot \binom{n}{k-1} < (1-\epsilon)\ln n$ for some $\epsilon > 0$, then a.a.s. $\mathcal{H}(n,p,k)$ is disconnected. On the other hand, if $p \cdot \binom{n}{k-1} > (1+\epsilon)\ln n$ for some $\epsilon > 0$, then a.a.s. $\mathcal{H}(n,p,k)$ is connected.

Finally, let us present a very simple experiment which demonstrates that we indeed lose some important information by restricting ourselves to 2-section graphs. Consider the three hypergraphs depicted on Figure 7.1 which have the same 2-section graphs. Suppose that the occurrence of, say, H_2 in some large hypergraph H indicates some anomalous behaviour whereas seeing H_1 or H_3 is perfectly normal. A natural approach to identifying anomalous sets of four nodes would be to inspect the 2-section of H using standard tools for graphs, finding all sets inducing graph H_1 in $[H]_2$, and then investigating these potentially anomalous sets after coming back to hypergraphs. Unfortunately, it might be the case that one would find a large number of sets of potential anomalies in $[H]_2$ but none of them actually are.

In order to illustrate this potential problem, we consider hypergraphs on $n = 500$ nodes. We fix $p_2 = c/(n-1)$, $p_3 = (8-c)/((n-1)(n-2))$ for some real number $c \in [0,8]$, and with all other p_is equal to zero. This choice yields constant expected degree in the corresponding 2-section graphs (with parallel edges allowed). Indeed, since each hyperedge of size 3 consisting of node v in the hypergraph yields two edges adjacent to v in the 2-section graph, it follows from (7.1) that the expected degree of any node in the 2-section graph is equal to

$$\binom{n-1}{1}p_2 + 2 \cdot \binom{n-1}{2}p_3 = c + (8-c) = 8.$$

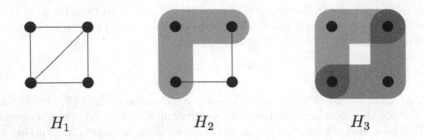

$$H_1 \qquad\qquad H_2 \qquad\qquad H_3$$

FIGURE 7.1
These three hypergraphs have the same 2-section, which is precisely H_1, but which appear in $\mathcal{H}(n, \mathcal{P})$ with different probabilities.

TABLE 7.2
Average number of patterns depicted in Figure 7.1 in random hypergraphs of size $n = 500$ with $p_2 = c/(n-1)$, $p_3 = (8-c)/((n-1)(n-2))$.

c	H1	H2	H3
0	0.0000	0.0000	15.4375
1	0.0000	2.1250	11.5000
2	0.0000	11.9375	7.1875
3	0.1250	21.6875	6.8125
4	0.4375	31.4375	3.5626
5	1.5000	37.9375	1.7500
6	3.5000	35.5625	1.4375
7	9.0000	24.0000	0.3750
8	15.0625	0.0000	0.0000

For each value $c \in \{0, 1, \dots, 8\}$, we generated 16 random hypergraphs and we report the average number of sets of four nodes giving hypergraph motifs H_1, H_2 and H_3 in Table 7.2. As expected, for small and large values of c, there are a lot of potentially anomalous sets (based on the 2-section "footprint") but very few of them are actually anomalous. For such scenarios, it is advised to stay with hypergraphs and try to detect motifs there. On the other hand, for average values of c (say, $c = 5$) almost all potentially anomalous sets actually are anomalous so moving to the 2-section might not be a bad idea.

Finally, let us mention that from the computational point of view, in this specific example finding a H_2 motif in a hypergraph should be a computationally easier task than finding a H_1 motif in the corresponding 2-section graph. However, in general, whether it is faster to look for a given motif in a hypergraph or in the corresponding 2-section graph depends on the motif and implementation of the algorithm used to search for it.

Another reason to introduce random models we identified in Section 2.1 is to create a synthetic graph that closely resembles a real-world network (reason (iii)). This is typically done in order to create a flexible laboratory for testing various scenarios. Examples of such models include the **LFR** and **ABCD** graphs we discussed in Section 5.3. In the context of community detection in hypergraphs, we will be interested in modelling various levels of heterogeneity of hyperedges. These synthetic networks with an engineered **ground truth** will be used to evaluate the performance of various clustering algorithms. As we aim for a simple model and because the degree distribution should not affect our exploratory experiments, we propose a model that is inspired by the classical **stochastic block model** that we also discussed in Section 5.3. We do *not* claim the model is realistic; it is rather a "toy-model" for illustration purpose.

Let $\mu \in [0,1]$ and $\tau \in (0.5, 1]$. The two variants of the **hypergraph with community structure**, $\mathcal{H}_=(\mu, \tau)$ and $\mathcal{H}_\geq(\mu, \tau)$, generate random hypergraphs with specified community sizes and a given distribution of hyperedge sizes. The parameter μ controls the fraction of edges that are between communities, and parameter τ models various levels of homogeneity of hyperedges.

A random hypergraph $\mathcal{H}_{...}(\mu, \tau)$ consists of K communities; the kth community has n_k members so the total number of nodes in $\mathcal{H}_{...}(\mu, \tau)$ is equal to $n = \sum_{k=1}^{K} n_k$. For $2 \leq d \leq M$, m_d is the number of hyperedges of size d; in particular, M is the size of a largest hyperedge and $m = \sum_{d \geq 2} m_d$ is the total number of hyperedges. Hyperedges are partitioned into *community* and *noise* hyperedges. The expected proportion of noise edges is $\mu \in [0,1]$, the parameter that controls the *level of noise*. Each community hyperedge will be assigned to one community. The expected fraction of hyperedges that are assigned to the kth community is p_k; in particular, $\sum_{k=1}^{K} p_k = 1$. Community hyperedges that are assigned to the kth community will primarily consist of members of that community. On the other hand, noise hyperedges will be "sprinkled" across the whole hypergraph.

The hyperedges of $\mathcal{H}_{...}(\mu, \tau)$ are generated as follows. For each edge size d, we independently generate m_d edges of size d. For each edge e of size d, we first decide if e is a community hyperedge or a noise hyperedge. It is a noise with probability μ; otherwise, it is a community hyperedge. If e turns out to be a noise hyperedge, then we simply choose its d nodes uniformly at random from the set of all sets of nodes of size d, regardless of which community they belong to. On the other hand, if e is a community edge, then we assign it to community k with probability p_k. Then, we fix the homogeneity value τ_e of hyperedge e based on the parameter $\tau \in (0.5, 1]$ of the model. There are two variants of the model and the way the value of τ_e is generated for a given edge e is the only difference between the two variants:

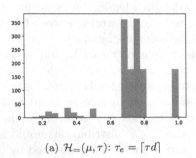
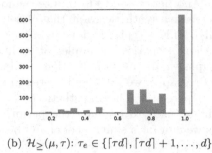

(a) $\mathcal{H}_=(\mu, \tau)$: $\tau_e = \lceil \tau d \rceil$ (b) $\mathcal{H}_\geq(\mu, \tau)$: $\tau_e \in \{\lceil \tau d \rceil, \lceil \tau d \rceil + 1, \ldots, d\}$

FIGURE 7.3

Comparing two different ways of assigning community edge homogeneity τ_e.

- in $\mathcal{H}_=(\mu, \tau)$, $\tau_e = \lceil \tau d \rceil$ (deterministically); in this case, τ represents the *homogeneity* of community edge e,

- in $\mathcal{H}_\geq(\mu, \tau)$, τ_e is chosen uniformly at random from the set $\{\lceil \tau d \rceil, \lceil \tau d \rceil + 1, \ldots, d\}$; this time τ represents the *lower bound homogeneity* of community edge e.

Finally, members of e are determined as follows: τ_e nodes are selected uniformly at random from the kth community, and the remaining nodes are selected uniformly at random from nodes outside of this community. We compare the two variants of selecting τ_e in Figure 7.3 by performing the following experiment. Both hypergraphs, $\mathcal{H}_=(\mu, \tau)$ and $\mathcal{H}_\geq(\mu, \tau)$, consist of $n = 1,000$ nodes and $m = 1,400$ hyperedges, 200 each of sizes 2 to 8. Moreover, the number of communities is $K = 10$, the level of noise is $\mu = 0.1$, and $\tau = 0.65$. In particular, we see that $\mathcal{H}_\geq(\mu, \tau)$ leads to much more homogeneous community edges in comparison to $\mathcal{H}_=(\mu, \tau)$.

As mentioned above, the proposed model is aimed to be simple but it tries to capture the fact that many real-world networks represented as hypergraphs exhibit various levels of homogeneity or the lack of thereof. Moreover, some networks are noisy with some fraction of hyperedges consisting of nodes from different communities. Such behaviour can be controlled by parameters $\tau_{\min}$ and μ. Thus, we have a tool to test the performance of our algorithms for various scenarios. A good algorithm should be able to adjust to any scenario in an unsupervised way.

Yet another reason for introducing random models that we discussed in Section 2.1 is to use them to guide the optimization process of some algorithm of interest (reason (iv)). For example, we used the **Chung-Lu** random graph model (see Section 2.5) as the **null model** in order to define the **modularity function** that is then used by the **Louvain** or **ECG** algorithms to detect communities (see Section 5.4). In order to adjust these ideas and to define hypergraph modularity for finding communities in hypergraphs, we will use the generalization of the **Chung-Lu** model to hypergraphs.

Consider a given hypergraph $H = (V, E)$ with $V = [n]$, where hyperedges $e \in E$ are subsets of V with cardinality greater than one (we may ignore hyperedges of size one as they do not affect the modularity function). Since we are concerned with hypergraphs that are not necessarily simple, hyperedges are multi-sets. Such hyperedges can be described using distinct sets of pairs $e = \{(i, m_e(i)) : i \in V = [n]\}$ where $m_e(i) \in \mathbb{N} \cup \{0\}$ is the multiplicity of the node i in e (including zero which indicates that i is not present in e). Then $|e| = \sum_i m_e(i)$ is the size of hyperedge e and the degree of a node i in H is $\deg_H(i) = \sum_{e \in E} m_e(i)$. When the reference to the hyperedge is clear from the context, we simply use m_i to denote $m_e(i)$. As for graphs, the volume of a subset $A \subseteq V$ is $\text{vol}_H(A) = \sum_{v \in A} \deg_H(v)$.

Similarly to what was done for graphs, we define a random model on hypergraphs, $\mathcal{H}(H)$, where the expected degrees of all nodes are the corresponding degrees in H. To simplify the notation, we omit explicit references to H in the remaining of this section; in particular, $\deg(v)$ denotes $\deg_H(v)$, $\mathcal{H}$ denotes $\mathcal{H}(H)$, E_d denotes the edges of H of size d and $H_d = (V, E_d)$. Moreover, we use E' to denote the edge set of H'. Finally, let F_d be the family of multi-sets of size d; that is,

$$F_d := \left\{ \{(i, m_i) : i \in [n]\} : \sum_{i=1}^n m_i = d \right\}.$$

Now, we are ready to define a generalization of the **Chung-Lu** model to hypergraphs.

Let H be any hypergraph on n nodes. We define $\mathcal{H} = \mathcal{H}(H) = ([n], E)$ to be the probability distribution of hypergraphs (including non-simple hypergraphs) on the set of nodes $[n]$. The hyperedges are generated via independent random experiments. For each d such that $|E_d| > 0$, the probability of generating the edge $e \in F_d$ is given by:

$$P_{\mathcal{H}}(e) = |E_d| \cdot \binom{d}{m_1, \ldots, m_n} \prod_{i=1}^n \left(\frac{\deg(i)}{\text{vol}(V)} \right)^{m_i}. \tag{7.3}$$

Let $(X_1^{(d)}, \ldots, X_n^{(d)})$ be the random vector following a multinomial distribution with parameters $d, p_{\mathcal{H}}(1), \ldots, p_{\mathcal{H}}(n)$, where $p_{\mathcal{H}}(i) = \deg(i)/\text{vol}(V)$ and $\sum_{i \in [n]} p_{\mathcal{H}}(i) = 1$; that is,

$$s_{\mathcal{H}}(e) := \mathbb{P}\left((X_1^{(d)}, \ldots, X_n^{(d)}) = (m_1, \ldots m_n) \right) = \binom{d}{m_1, \ldots, m_n} \prod_{i=1}^n (p_{\mathcal{H}}(i))^{m_i}.$$

Note that this is the expression found in (7.3); that is, $P_{\mathcal{H}}(e) = |E_d| \cdot s_{\mathcal{H}}(e)$. As a result, alternatively one can think about the following auxiliary process. Select a random multi-set consisting of d nodes (counting possible repetitions); in d independent rounds, node v_i is selected with probability $p_{\mathcal{H}}(i)$. Repeat this experiment $|E_d|$ times and use the expected number of times edge e occurred in this process for the value of $P_{\mathcal{H}}(e)$. An immediate consequence of this coupling between the two processes is that the expected number of edges of size d is $|E_d|$. Finally, as with the original **Chung-Lu** (graph) model, if $P_{\mathcal{H}}(e) > 1$, then it should be regarded as the expectation and a multi-hypergraph should be considered instead. However, as before, from the practical point of view it is safe to assume that all $P_{\mathcal{H}}(e) \leq 1$.

In order to compute the expected d-degree of a node $i \in V$ of some hypergraph H', note that

$$\deg_{H'_d}(i) = \sum_{e \in F_d} m_e(i) \cdot \mathbb{I}_{\{e \in E'\}},$$

where $\mathbb{I}_{\{\}}$ is the indicator random variable for the corresponding event. Hence, using the linearity of expectation, then splitting the sum into $d + 1$ partial sums for different multiplicities of i, we get:

$$\mathbb{E}_{H' \sim \mathcal{H}}\left(\deg_{H'_d}(i) \right) = \sum_{e \in F_d} m_e(i) \cdot P_{\mathcal{H}}(e) = |E_d| \sum_{e \in F_d} m_e(i) \cdot s_{\mathcal{H}}(e)$$

$$= |E_d| \sum_{m=0}^{d} m \cdot \sum_{e \in F_d; m_e(i)=m} s_{\mathcal{H}}(e)$$

$$= |E_d| \sum_{m=0}^{d} m \cdot \mathbb{P}(X_i^{(d)} = m)$$

$$= |E_d| \sum_{m=0}^{d} m \cdot \binom{d}{m} (p_{\mathcal{H}}(i))^m (1 - p_{\mathcal{H}}(i))^{d-m}$$

$$= |E_d| \cdot d \cdot p_{\mathcal{H}}(i).$$

The second last equality follows from the fact that we obtained the expected value of a random variable with binomial distribution. One can compute the expected degree as follows:

$$\mathbb{E}_{H' \sim \mathcal{H}}[\deg_{H'}(i)] = \sum_{d \geq 2} \frac{d \cdot |E_d| \cdot \deg(i)}{\mathrm{vol}(V)} = \deg(i),$$

since $\mathrm{vol}(V) = \sum_{d \geq 2} d \cdot |E_d|$. Hence, indeed, the model produces a random hypergraph $\mathcal{H}$ with an expected degree distribution that is equal to the degree distribution of H.

7.4 Community Detection in Hypergraphs

There have been some attempts recently to deal with hypergraphs in the context of clustering. Unfortunately, there are many ways such extensions can be done depending on how often nodes in one community share hyperedges with nodes from other communities. This is something that varies between networks at hand and usually depends on the sizes of the hyperedges. Indeed, in the collaboration hypergraph we discussed earlier, hyperedges associated with papers written by mathematicians might be more homogeneous and smaller in comparison than with those written by medical doctors who tend to work in large and multidisciplinary teams. Moreover, in general, papers with a large number of co-authors tend to be less homogeneous. A good algorithm should be able to automatically decide which extension should be used.

Let us generalize the modularity function that we introduced in Section 5.4 for graphs to hypergraphs. For hyperedges of size greater than 2, several definitions can be used to quantify the edge contribution of a given partition $\mathcal{A} = \{A_1, A_2, \ldots, A_\ell\}$ of the set of nodes. As a result, the choice of the hypergraph modularity function is not unique. The choice is dependent upon how strongly one believes that a hyperedge is an indicator that some of its nodes fall into one community. The fraction of nodes of a given hyperedge that belong to one community is called its *homogeneity* (provided it is more than 50%). In one extreme case, all nodes of a hyperedge have to belong to one of the parts in order to contribute to the modularity function; this is the *strict* variant assuming that only the most homogeneous hyperedges provide information about underlying community structure. In the other natural extreme variant, the *majority* variant, one assumes that edges are not necessarily homogeneous and so a hyperedge contributes to one of the parts if more than 50% of its nodes belong to it; in this case being over 50% is the only information that is considered relevant for community detection. All variants in between guarantee that hyperedges contribute to at most one part. Alternatively, a hyperedge could contribute to the part that corresponds to the largest fraction of nodes. However, this might not uniquely determine the part and it is more natural to classify such edges as "noise" that should not contribute to any part anyway. Once the variant is fixed, one needs to benchmark the corresponding edge contribution using the degree tax computed for the generalization of the **Chung-Lu** model to hypergraphs that we proposed in Section 7.3.

In order to unify all potentially useful definitions of modularity functions, we put them into one common framework. This general framework is flexible and can be tuned and applied to hypergraphs with hyperedges of varying homogeneity. In order to achieve our goal, we "dissect" the modularity function so that each "slice" can be considered independently. For each hyperedge of size d, we will independently deal with contribution to the modularity function coming from hyperedges of size d with precisely c members from one of the parts, where $c > d/2$. For example, for $d = 7$ we get 4 slices corresponding to various values of c, namely, $c \in \{4, 5, 6, 7\}$.

Let $H = (V, E)$ be a hypergraph on the set of nodes $V = \{v_1, v_2, \ldots, v_n\}$ and the degree distribution $\mathbf{d} = (\deg(v_1), \deg(v_2), \ldots, \deg(v_n))$. Let $\mathcal{A} = \{A_1, A_2, \ldots, A_\ell\}$ be any partition of V. The contribution $q_H^{c,d}(\mathbf{A})$ to the modularity function coming from the slice corresponding to a given pair (c, d), $d/2 < c \leq d$, can be computed using the **generalized Chung-Lu** model as the **null-model**. As for graphs, we try to capture the discrepancy between the number of hyperedges of H of size d with exactly c members in some part of $\mathcal{A}$ and the corresponding expected value based on the **generalized Chung-Lu** model $\mathcal{H} = \mathcal{H}(H)$. It follows that

$$
\begin{aligned}
q_H^{c,d}(\mathbf{A}) &= \frac{1}{|E|} \sum_{A_i \in \mathbf{A}} \left(e_H^{d,c}(A_i) - \mathbb{E}_{H' \sim \mathcal{H}}[e_H'^{d,c}(A_i)] \right) \\
&= \frac{1}{|E|} \sum_{A_i \in \mathbf{A}} \left(e_H^{d,c}(A_i) - |E_d| \cdot \mathbb{P}\left(\mathrm{Bin}\left(d, \frac{\mathrm{vol}(A_i)}{\mathrm{vol}(V)} \right) = c \right) \right),
\end{aligned}
$$

where $e_H^{d,c}(A_i)$ is the number of hyperedges of size d that have exactly c members in A_i.

In order to unify the definitions, the modularity function for hypergraphs is controlled by *hyper-parameters* $w_{c,d} \in [0, 1]$ ($d \geq 2$, $\lfloor d/2 \rfloor + 1 \leq c \leq d$). For a fixed set of hyper-parameters, we simply define

$$
q_H(\mathbf{A}) = \sum_{d \geq 2} \sum_{c = \lfloor d/2 \rfloor + 1}^{d} w_{c,d} \, q_H^{c,d}(\mathbf{A}). \tag{7.4}
$$

This definition gives us more flexibility and allows us to value some slices more than others. Indeed, the majority variant can be considered as follows:

$$
q_H^m(\mathbf{A}) = \sum_{d \geq 2} \sum_{c = \lfloor d/2 \rfloor + 1}^{d} q_H^{c,d}(\mathbf{A}),
$$

that is, all hyper-parameters in (7.4) are equal to one. On the other extreme, the strict variant can be obtained as follows:

$$
q_H^s(\mathbf{A}) = \sum_{d \geq 2} q_H^{d,d}(\mathbf{A}),
$$

that is, only the slices with $c = d$ are considered.

In order to demonstrate the potential added benefit of staying with hypergraphs instead of reducing the problem to graphs we performed the following experiment. We investigated random hypergraphs with community structure $\mathcal{H}_=(\mu, \tau)$ we introduced in Section 7.3 with $n = 1,000$ nodes that form 10

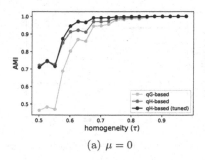
(a) $\mu = 0$

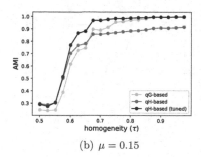
(b) $\mu = 0.15$

FIGURE 7.4
Average **AMI** over 16 random hypergraphs with communities with edge homogeneity parameter $\tau \in (0.5, 1]$. We show the results (a) without pure noise edges ($\mu = 0$) as well as (b) with noise edges ($\mu = 0.15$).

communities of size 100, and contain 200 hyperedges of size d for $2 \leq d \leq 8$; hence 1,400 random hyperedges were generated for each hypergraph in total. In Figure 7.4, we show experiments on hypergraphs $\mathcal{H}_=(0, \tau)$ and $\mathcal{H}_=(0.15, \tau)$ with two different levels of noise $\mu = 0$ and $\mu = 0.15$, and with $0.5 < \tau \leq 1$. For each choice of τ, we independently generated 16 random hypergraphs.

For each of them, we run the classic **Louvain** algorithm on the corresponding 2-section graphs (see Section 5.4), some ad-hoc and simple "Louvain type" algorithm that uses the hypergraph modularity function (7.4) as well as a variation with hyper-parameters tuned to reflect the homogeneity of tested random hypergraphs. We measured the performance of these algorithms using the **Adjusted Mutual Information** (**AMI**) score we introduced in Section 5.3. In short, the **AMI** is the information theory measure that allows us to quantify the similarity between two partitions of the same set of nodes, the partition returned by the algorithm and the ground truth.

In Figure 7.4(a), we show the results for $\mathcal{H}_=(0, \tau)$, so the only source of noise is present in the form of non-homogeneous community edges. In this case, taking the average **AMI** over 16 hypergraphs for each choice of τ, we see that the algorithm based on hypergraph modularity (qH) performs better; it is particularly effective when the true value τ is known (tuned version). In Figure 7.4(b), we see that the result is slightly different for $\mathcal{H}_=(0.15, \tau)$ in the presence of noise edges. This time, the qH-based algorithm performs better for small values of τ, but the **Louvain** (qG-based) algorithm performs better as we increase τ, unless we are given its true value (tuned version). This is still experimental but the general tendency we observe is that using hypergraph-based algorithms is advantageous when a large proportion of the noise in the communities comes from the non-homogeneity of edges rather than noise edges.

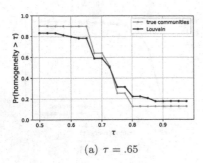
(a) $\tau = .65$

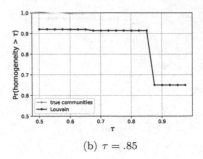
(b) $\tau = .85$

FIGURE 7.5
Estimating the homogeneity parameter τ via the clustering performed on the 2-section graph (Louvain algorithm).

The previous experiment shows that knowing some global statistics (namely, how homogeneous the network is) significantly increases the performance of our prototype algorithms. However, typically such information is not available and the algorithm has to learn such global statistics in an unsupervised way. In our second experiment, we test if this is possible. We take a partition returned by the **Louvain** algorithm run on the corresponding 2-section graphs and investigate all hyperedges of $\mathcal{H}$. For each hyperedge e we check if at least $\tau \geq 0.5$ fraction of its nodes belong to some community. We compare it with the corresponding homogeneity value based on the *ground truth*. The two distributions are presented in Figure 7.5 for two hypergrahs respectively with $\tau = 0.65$ and $\tau = 0.85$, $\mu = 0.1$ and with the same parameters as in the previous experiment. We see that the distributions are very close, which suggests that learning the right value of τ should be possible in practice. These initial experiments are very encouraging but clearly more work needs to be done.

7.5 Experiments

For this chapter's experiments, we built a hypergraph by considering scenes from the television series **Game of Thrones** with data obtained from GitHub[1]. In this hypergraph, the nodes correspond to characters in the series, and hyperedges are groups of characters appearing in the same scene(s). Finally, hyperedge weights are total scene(s) duration in seconds involving those characters. We kept hyperedges with at least 2 characters and we recursively discarded characters with degree below 5 to focus on the better known characters. This procedure can be viewed as a natural generalization of the k-core

---

[1]github.com/jeffreylancaster/game-of-thrones

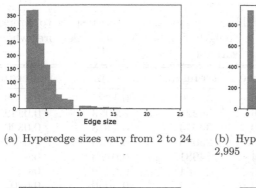

(a) Hyperedge sizes vary from 2 to 24

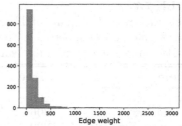

(b) Hyperedge weights vary from 2 to 2,995

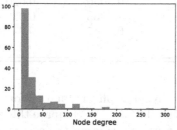

(c) Node degrees vary from 5 to 306

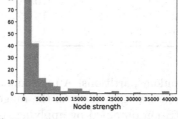

(d) Node strength vary from 102 to 39,899

FIGURE 7.6
Descriptive statistics for the Game of Thrones hypergraph.

algorithm for graphs we discussed in Section 3.5. This yields a hypergraph with 173 nodes of degrees between 5 and 306, 1,432 hyperedges of size between 2 and 24, and weights ranging from 2 to 2,995. For the nodes, we also define the **strength** as the sum of the weights of the edges that contain this node; this amounts to the total scene duration for each character (excluding the scenes that were pruned to build this hypergraph). We summarize the main characteristics of this hypergraph in Figure 7.6.

For comparison purposes, we also built the corresponding 2-section graph representation of this hypergraph. In particular, we used it to compute a few centrality measures we saw earlier in this book: PageRank and betweenness. In Table 7.7, we show these values for the top-10 nodes with respect to strength. We see that degree, strength, and PageRank scores are highly correlated. However, perhaps surprisingly, this is not the case for betweenness; some characters have higher values which is indicative of interaction with several different groups of characters throughout the series.

Next, we look at hypergraph modularity with linear weights $w_{c,d} = c/d$ (see 7.4) over random partitions of the nodes. We considered random partitions between 2 and 20 parts (inclusively), and for each choice we generated

TABLE 7.7

Top characters in the Game of Thrones hypergraph with respect to strength. We see high correlation between strength, degree and pagerank scores whereas the betweenness scores are more variable.

name	degree	strength	betweenness	pagerank
Tyrion Lannister	263	39899	0.1027	0.0408
Jon Snow	306	39221	0.0023	0.0442
Daenerys Targaryen	220	30644	0.1178	0.0320
Sansa Stark	183	25009	0.1065	0.0277
Cersei Lannister	184	24981	0.0195	0.0265
Jaime Lannister	153	22741	0.0419	0.0252
Jorah Mormont	118	19344	0.0050	0.0203
Arya Stark	103	17775	0.1753	0.0235
Davos Seaworth	131	16960	0.0547	0.0189
Lord Varys	115	15615	0.0660	0.0160

10 random partitions. After computing the hypergraph modularity for each partition, we got values in the range $[-0.0786, 0.0447]$. We compare this with a partition obtained by applying the **Louvain** algorithm on the corresponding 2-section graph, for which we got the hypergraph modularity $qH = 0.526$, much higher that any of the random partitions. This shows that the situation with hypergraphs is similar to the one we experienced with graphs. Almost all partitions have very low modularity and so our (challenging) task is to find the "needle in the haystack." As a result, it is important to carefully design an efficient algorithm that searches the space of all partitions and which finds a good partition without hoping for the best one.

The partition with large hypergraph modularity is shown in Figure 7.8 where we plot the corresponding 2-section graph. The tight cluster we see at the bottom left corresponds to the characters from the Theatre Troupe seen in the series. We also ran the same "ad-hoc" Louvain-based hypergraph clustering algorithm we used in Section 7.4 on this graph with little difference in the clusters, at least with respect to the main characters.

Finally, we consider one of the main characters of the series, Daenerys Targaryen, and we look at the other characters present in the same cluster. In Table 7.9, we list the top characters in that cluster with respect to the corresponding node's strength. Readers familiar with the series will recognize those characters as having lots of common scenes.

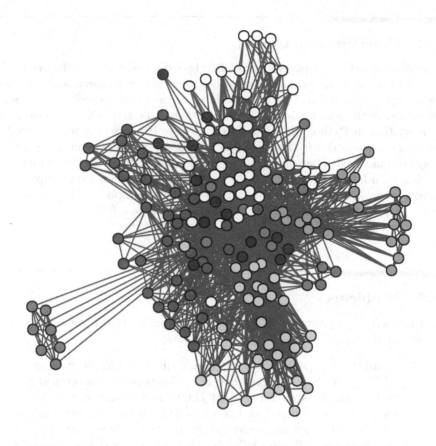

FIGURE 7.8
Clusters for the Game of Thrones hypergraph represented as the 2-section graph. The tight sub-cluster at the bottom left corresponds to the Braavos Theatre Troupe.

TABLE 7.9
Top characters in the Game of Thrones hypergraph with respect to strength in the same cluster as main character Daenerys Targaryen.

name	degree	strength	betweenness	pagerank
Daenerys Targaryen	220	30644	0.1178	0.0320
Jorah Mormont	118	19344	0.0050	0.0203
Missandei	92	13683	0.0025	0.0139
Grey Worm	79	10416	0.0152	0.0110
Barristan Selmy	35	6514	0.0000	0.0072

7.6 Practitioner's Corner

The development of hypergraph-specific algorithms as well as software packages to handle hypergraphs is currently a very active research and development topic. As a result, it is impossible to recommend a specific package at this time. In the accompanying notebook, we use the HyperNetX package[2] as it is written in Python and it is simple to use. If one wants to use the Julia language to work with hypergraphs, the `SimpleHypergraphs.jl`[3] package may be considered. There are many recent developments of specific algorithms to deal with hypergraphs such as the author's own work on hypergraph modularity and clustering, or generalization of several graph-based measures to hypergraphs via the use of high-order walks, to name a few.

7.7 Problems

In this section we present a collection of potential practical problems for the reader to attempt.

1. Consider $\mathcal{H}(n, p, 3)$, random 3-uniform hypergraph with $p = c/\binom{n-1}{2}$ for some constant $c \in \mathbb{R}_+$. Compare the theoretical prediction for the size of the giant component (equation (7.2)) with empirical results based on 1,000 independent runs for small graphs on $n = 100$ nodes and larger graphs on $n = 10,000$ nodes. Present the results on a figure similar to what we did for $\mathcal{G}(n, p)$, binomial random graphs—Figure 2.1. In order to find the size of the giant component, you might simply do it on the corresponding 2-section graph.

2. For the Game of Thrones (weighted) hypergraph, use HyperNetX library to do the following.

   a. Compute the degree and the strength of each node.

   b. Compute the correlation between those two quantities (degree and strength): **Pearson's** ρ, **Spearman's** r_s, and **Kendall's** τ discussed in Section 3.4.

   c. Fit the regression line predicting strength from degree; find its slope, plot the line and all points $(\deg(v), \text{strength}(v))$.

   d. Find top 5 characters with the largest degree and top 5 characters with the largest strength.

---

[2] github.com/pnnl/HyperNetX
[3] github.com/pszufe/SimpleHypergraphs.jl

3. Let us start with a few definitions. For a given $s \in \mathbb{N}$, **s-edge-walk** is a sequence of edges where successive pairs share at least s nodes. Similarly, **s-node-walk** is a sequence of nodes where successive pairs share at least s edges. Finally, s-connected components (edge or node) have s-walk between every pair (edge or node, respectively). In the following, you can use the HyperNetX function `s_components()` with the appropriate parameters to get the s-connected components.

   For the Game of Thrones (unweighted) hypergraph, use HyperNetX library to solve the following questions related to node connected components.

   a. Plot the number of s-connected components for all values of s until all nodes are disjoint.

   b. For each unique partition obtained in the previous question for a given value of s, compute the (hypergraph) modularity function q_H. Which s-value yields the best q_H?

   c. What is the maximum value of s for which there are still non-trivial components (that is, components of size at least 2)? What characters are in those components?

   d. What characters are in non-trivial 50-connected components?

   For the Game of Thrones (unweighted) hypergraph, use HyperNetX library to solve the following questions related to edge connected components.

   e. Plot the number of s-connected components for all values of s until all edges are disjoint.

   f. What is the maximum value of s for which there are still non-trivial components (that is, components of size at least 2)? What characters are in ALL edges in those components? What characters are in at lest 1 edge in those components?

4. Let us start with a definition. For a given $s \in \mathbb{N}$, provided that the graph is s-connected, the **s-diameter** is the length of a longest s-node-path between two nodes. The `diameter()` and `distance()` functions are available in HyperNetX.

   For the Game of Thrones (unweighted) hypergraph, use HyperNetX library to do the following.

   a. Find the s-diameter for $s = 1$ and $s = 2$? What about $s = 3$?

   b. Sample 200 random pairs of nodes from the hypergraph and compare s-distance distribution for $s = 1$ and $s = 2$.

7.8   Recommended Supplementary Reading

- S.G. Aksoy, C.A. Joslyn, C.M. Ortiz Marrero, B.L. Praggastis, E. Purvine, "Hypernetwork Science via High-Order Hypergraph Walks", EPJ Data Science 9, no. 1:(16). PNNL-SA-144766.

- B. Kamiski, V. Poulin, P. Praat, P. Szufel, F. Thberge, "Clustering via hypergraph modularity", *PLoS ONE* 14(11): e0224307 (2019).

- B. Kamiski, P. Praat, F. Thberge, "Community Detection Algorithm Using Hypergraph Modularity" in: *Complex Networks & Their Applications IX. COMPLEX NETWORKS 2020*. Studies in Computational Intelligence, vol 943. Springer, Cham.

  Survey on the higher-order architecture of real complex system:

- F. Battiston, G. Cencetti, I. Iacopini, V. Latora, M. Lucas, A. Patania, J.-G. Young, G. Petri. "Networks beyond pairwise interactions: structure and dynamics", *Physics Reports*, 2020.

Part II

Additional Material

8

Detecting Overlapping Communities

Graph clustering is a well-studied graph mining problem. It is particularly relevant in cases where the set of nodes is partitioned into non-overlapping communities. In Chapter 5, we already saw a number of algorithms for community detection such as **Louvain** and **ECG**. In this chapter, we revisit the problem of graph clustering and generalize it to include situations in which:

- nodes can be part of several communities (overlapping communities);

- nodes can be part of no community ("noise" nodes).

There are different ways to approach this problem, including:

1. methods based on finding overlapping cliques (recall that a clique is another name for a complete subgraph),

2. methods based on splitting nodes into multiple personae, and

3. methods based on clustering edges instead of nodes.

We illustrate some of those methods using the graph of Zachary's Karate Club which we experimented with earlier in the book as well as a few other graphs.

8.1 Overlapping Cliques

The first algorithm we consider is a **Clique Percolation Method (CPM)**. For a fixed clique size k (typically k is set to be 3 or 4), the algorithm works as follows. For a given graph G, we create a new graph G' whose nodes are all k-cliques of the original graph G. An edge between two nodes in G' is formed if the corresponding k-cliques in the original graph G have $k-1$ nodes in common. Once G' is formed, we find connected components in G' that partition the set of nodes of G'. Nodes of the original graph G that belong to at least one clique of one such connected component form a community. Note that, in particular, a node in G may become a member of two communities (if it is contained in two k-cliques that belong to two connected components in G') or can be a member of no community (if it is not contained in any k-clique).

DOI: 10.1201/9781003218869-8

For weighted graphs, one can obtain a hierarchy of communities using **CPM** by pruning edges with weight below some variable threshold. When all edges are pruned, all nodes are noise nodes, that is, they belong to no community which can alternatively be viewed as them only belonging to their own communities. With more edges kept, new cliques in G' may be formed and some communities may be merged. In order to get a hierarchy of communities for unweighted graphs, one can use the **ECG**-generated edge weights (votes) as follows:

1. run **ECG** to get weights (number of votes) for each edge,

2. rank the edge weights from the largest (many votes, strong association) to the smallest (few or no votes, weak association),

3. for a given threshold value, drop edges below that value and run **CPM**.

We run the **CPM** algorithm "as is" on the Zachary graph with $k = 3$ and show the results in the left plot of Figure 8.1. Nodes that belong to two or more clusters are represented as squares. We obtained one large community, two small ones, and two orphan nodes (shown in white). All graphs shown here can be seen more clearly in colour in the accompanying notebook.

Let us now move to the hierarchical approach. **ECG** induces edge weights from a discrete list of values since we use a fixed-size ensemble. As a result, we may use each possible weight as a threshold in decreasing order, pruning edges with weight less than or equal to that threshold. This yields a hierarchy of overlapping clusters, which do not necessarily cover all the nodes. In the right plot of Figure 8.1, we show the penultimate result (all edges with at least one vote are kept), where pruned edges are shown in pale grey. In this case, we obtained two large communities, two small ones and three isolated nodes (shown in white). The complete hierarchy of graphs is generated in the accompanying notebook.

One can also look at the hierarchy of **CPM** communities from the perspective of some given node(s) of interest; we show some examples in the accompanying notebook.

8.2   Ego-splitting

Let $G = (V, E)$ be any graph (weighted or unweighted) on n nodes. For a given node $v \in V$, we say that the ego-net of v is the graph induced by v and $N(v)$, the set of neighbours of v. In this section, we consider the following ego-net splitting framework:

1. For each node v:

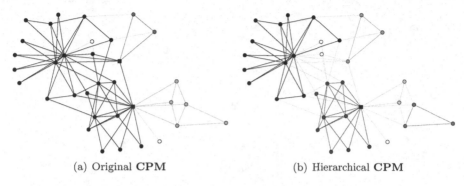

(a) Original **CPM**   (b) Hierarchical **CPM**

FIGURE 8.1
Results of the **CPM** algorithm on the Zachary graph with $k = 3$. The left plot (a) shows the resulting communities with the original version, and the right plot (b) shows the result when edges with no **ECG** vote are pruned.

(a) build the ego-net of v but excluding node v;

(b) partition this ego-net using some local method such as **label propagation** (**LP**; see Section 5.6) or simply by finding **connected components** (**CC**; see Section 1.7);

(c) split node v into multiple personae, one persona per ego-net cluster.

2. Cluster this new graph (with duplicated nodes) with some graph partitioning algorithm such as **LP**, **Louvain** or **ECG**.

3. Each node from the original graph is assigned to the communities that at least one of its duplicates belongs to. In order to avoid having many small communities, we can set a minimum community size. We use minimum size of 3 in what follows.

There are a few variations of the **LP** algorithm; for our experiments, we use the one implemented in `igraph`. We show the resulting communities in Figure 8.2 using **CC** to partition the ego-net, and **LP** for clustering; there are four communities and two isolated nodes. Note that, unlike **CPM**, the results may vary if one runs this algorithm multiple times since **LP** is a stochastic algorithm. More examples can be found in the accompanying notebook.

8.3   Edge Clustering

In this section, we obtain overlapping communities by clustering edges instead of nodes. The algorithm can be described as follows:

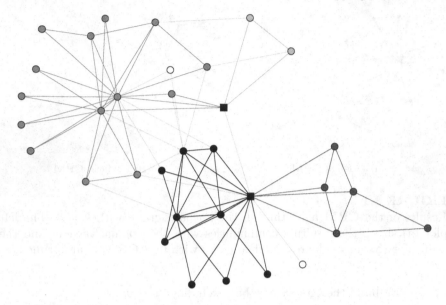

FIGURE 8.2
Ego-splitting of the Zachary graph with **CC** in step 1(b) and **LP** in step 2.

1. For each pair of edges sharing a node, say $v_i v_k$ and $v_j v_k$, compute some similarity measure between the neighbourhoods $N(v_i), N(v_j)$ of nodes v_i and v_j. For example, one may use the **Jaccard index** $|N(v_i) \cap N(v_j)|/|N(v_i) \cup N(v_j)|$.

2. Perform hierarchical clustering on the edges with this similarity measure. Nodes that are incident to at least one edge from a given community belong to that community. As before, we can impose a minimal number of nodes in a community.

In order to get a clustering of edges in step 2 above, it is convenient to reduce the problem to clustering of nodes by considering the line graph of G. For a given graph $G = (V, E)$, its **line graph** is a graph $G' = L(G) = (V', E')$ in which $V' = E$ and two nodes in G' (edges in G), $e_1, e_2 \in V' = E$, are adjacent if the corresponding edges in G share a node. Note that in our scenario, $L(G)$ is a weighted graph where weights are similarity measures of the associated nodes in $L(G)$ (edges in G).

Now, one may use some hierarchical clustering of nodes of $L(G)$, including those mentioned in Section 5.5. In our implementation, we simply consider the connected components, so the hierarchy can be obtained naturally by varying the threshold for the edge weights (**Jaccard** measures) and we select the best clustering in the hierarchy based on the **modularity** scores on the line graph (see Section 5.4). Examples are shown in the accompanying notebook.

8.4   Illustration: Word Association Graph

We consider the graph built from the free-association database[1]. In a nutshell, 5,017 base words were selected and, for each word, several thousand participants listed associated words, which were not necessarily base words. A total of 72,176 different words were chosen as responses with over 750,000 associations for the 5,017 base words.

We built an association graph on the set W of the base words. For each word $w_i \in W$, an association strength score is assigned for each other word $w_j \in W$ which equals the proportion of participants who selected w_j as an associated word given w_i. We denote this score as $s(w_j|w_i)$. Note that we usually have $s(w_i|w_j) \neq s(w_j|w_i)$ and so it is not symmetric. We built an undirected graph over all words in W by adding an edge between words w_i and w_j if $s(w_j|w_i) + s(w_i|w_j) \geq 0.025$. This gave us a graph with 29,266 edges. From this association graph, we can apply hierarchical **CPM** in two different ways: using the scores $s(w_j|w_i) + s(w_i|w_j)$ for the edge weights, or using the **ECG**-based weights. We looked at both, using $k = 4$ for the clique size. We see that finding overlapping clusters is a useful tool to capture different meanings or different usages of common words.

Consider the word MATH (all words are represented with capital letters in the database). Using **ECG**-based hierarchical **CPM**, here is an example of a set of clusters we get in the hierarchy:

{ALGEBRA,CALCULUS,NUMBER,FACTOR,TRIGONOMETRY,ARITHMETIC,MATH}
{MATH,ALGEBRA,FORMULA,EQUATION}
{COMPUTER,CALCULATE,COMPUTE,FIGURE,ADD,CALCULATOR,MATH}
{DIVISION,DIVIDE,ADD,SUBTRACT,MULTIPLY,QUOTIENT,MATH}

The results are quite similar using the association strength scores; here is a set of clusters from that hierarchy:

{ALGEBRA,CALCULUS,HARD,TRIGONOMETRY,MATH}
{CALCULATE,COMPUTE,FIGURE,ADD,MATH}
{MATH,ALGEBRA,FACTOR,NUMBER}
{DIVISION,DIVIDE,ADD,SUBTRACT,MULTIPLY,QUOTIENT,MATH}

In the notebook, we also provide graphical representations which are better viewed in colour. We also tried the same approach on other words. The word MONEY is particularly rich, with up to 20 different clusters in the hierarchy. We list a few of those here:

{PAY,MONEY,WAGE,SALARY}
{BILL,DUE,MONEY,DEBT,FEE,PAYMENT}
{NICKEL,QUARTER,MONEY,DOLLAR,CENT,PENNY,DIME,COIN,CENTS}
{EXPENSE,BILL,MONEY,COST,PAY,FEE,PRICE}

---

[1]http://w3.usf.edu/FreeAssociation/

8.5   Benchmark Graphs

Just like we did with partitioning algorithms (see Section 5.3), we compare some of the methods we presented here for overlapping clusters with a synthetic benchmark model. We use a simple generalization of the **LFR** benchmark[2] which allows us to specify that n' nodes out of n should be part of $c \geq 2$ communities. This is achieved via equipartition of the internal edges for nodes that are members of several communities. The other parameters, including the noise parameter μ, are the same as in the original **LFR** model.

In order to compare the (possibly) overlapping clusters with the ground truth, we use the following two measures:

- oNMI[3], a modification of **NMI**, which can handle overlapping communities. This can be done by generalizing cluster membership for a node to a random variable with some distribution over all possible clusters.

- **Omega index**[4], a generalization of **ARI**, which considers all pairs of nodes that appear together in k clusters for all possible values of $k \geq 0$.

First, let us explore if there is any bias with these measures as we saw earlier with the original non-adjusted measures. To do so, we generated an **LFR** graph with 1,000 nodes, 100 of which belong to 2 communities, and with the noise parameter set to $\mu = .25$. The graph has 25 ground-truth communities, and we generated random partitions with 2 to 60 communities, choosing 100 nodes at random to be part of 2 communities. We show the results in Figure 8.3(a). We see that the **oNMI** measure is almost flat at zero while the **Omega** index oscillates around zero, but not in any way that would suggest a bias. Moreover, as we will see next, the values are very small compared to the **Omega** index values we obtain with the different algorithms.

Next, we generate **LFR** graphs with the same parameters as the previous one (1,000 nodes, 100 that belong to 2 communities), but with noise parameter in the following range: $0.05 \leq \mu \leq 0.70$. For all algorithms, communities with less than 5 nodes are discarded. We show the results for the **oNMI** measure in Figure 8.3(c) and the **Omega** index in Figure 8.3(d) for three node-based algorithms we described: **CPM** (with 3-cliques), **ego-splitting** (with **CC** and **LP**), and the **Hierarchical CPM (H-CPM)** using **ECG** weights, for which we selected the penultimate level (keeping all edges that got at least one vote). From those plots, we see that dropping the edges with no **ECG** vote greatly helps with the **CPM** algorithm, and the **ego-splitting**

---

[2]sites.google.com/site/andrealancichinetti/files
[3]github.com/aaronmcdaid/Overlapping-NMI
[4]pypi.org/project/omega-index-py3

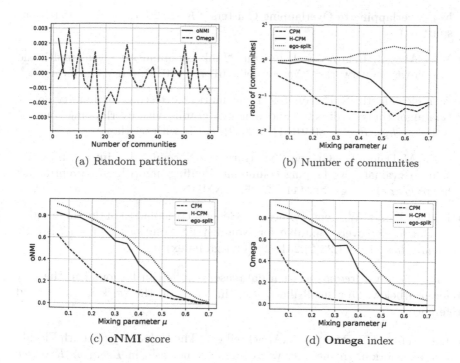

(a) Random partitions

(b) Number of communities

(c) oNMI score

(d) Omega index

FIGURE 8.3
Results in (a) with random partitions show no bias with the **oNMI** and **Omega** measures. In (b), we compare the number of communities obtained vs. the true number, with varying noise parameter μ. In (c), we compare various algorithms using the **oNMI** score and the **Omega** index, respectively.

algorithm perform quite well. In Figure 8.3(b), we look at the ratio between the number of communities found with each algorithm and the true number of communities, using a logarithmic scale. We see that **ego-splitting** tends to find more communities than the ground truth, while the opposite happens with **CPM**.

8.6 Recommended Supplementary Reading

- I. Derényi I.G. Palla, T. Vocsek, "Clique percolation in random networks", *Phys. Rev. Lett.*, 2005, vol. 94 pp. 160–202. (CPM)

- A. Epasto, S. Lattanzi, R. Paes Leme, "Ego-Splitting Framework: from

Non-Overlapping to Overlapping Clusters", *KDD* 2017, pp. 145–154. (ego-splitting)

- Y.Y. Ahn, J. Bagrow, S. Lehmann, "Link communities reveal multiscale complexity in networks", *Nature* 466, 761–764 (2010). (edge clustering)

- A. Lancichinetti, S. Fortunato, "Benchmarks for testing community detection algorithms on directed and weighted graphs with overlapping communities", *Phys. Rev. E* 80, 016118 (2009). (LFR with overlaps)

- A.F. McDaid, D. Greene, N. Hurley, "Normalized Mutual Information to evaluate overlapping community finding algorithms", preprint at: http://arxiv.org/abs/1110.2515. (oNMI)

- L.M. Collins, C.W. Dent, "Omega: A General Formulation of the Rand Index of Cluster Recovery Suitable for Non-disjoint Solutions", in *Multivariate Behav. Res.*, 1988 1;23(2):231–42. (Omega index)

For the word free association database, details can be found in the first reference below. Our small illustration in based on the work in the second reference.

- D.L. Nelson, C.L. McEvoy, T.A. Schreiber, "The University of South Florida free association, rhyme, and word fragment norms", in *Behavior Research Methods, Instruments and Computers* 2004, 36 (3), 402–407.

- G. Palla, I. Derényi, I.J. Farkas, T. Vicsek, "Uncovering the overlapping community structure of complex networks in nature and society", *Nature* 435(7043):814–818, July 2005.

9

Embedding Graphs

In this chapter, we illustrate a few applications of graph embedding on some chemical datasets. In this context, graph embedding refers to the embedding of a family of entire graphs, not the embedding of nodes of a single graph which we were concerned within Chapter 6.

9.1 NCI1 and NCI109 Datasets

NCI1 and NCI109 are two well-known datasets from the National Cancer Institute of chemical compounds that were screened for activity against non-small cell lung cancer and ovarian cancer cell lines, respectively[1]. The datasets are balanced, consisting of a similar number of inactive (label 0) and active (label 1) compounds. For each dataset, the compounds are represented as graphs with 37 and, respectively, 38 different node labels. In Table 9.1, we summarize the basic characteristics of the datasets in which each compound is represented as a graph made up of labelled nodes, and each graph also has a label (0 or 1). We show two of the graphs from the dataset in Figure 9.3, where we also indicate the node labels. In Figure 9.2, we plot the distribution of the number of nodes and, respectively, the number of edges for graphs with label 0 or 1 in the NCI109 dataset. The distribution for the NCI1 dataset is very similar. We immediately see that graphs with label 1 tend to have slightly more nodes and edges.

TABLE 9.1

Basic statistics of the two NCI datasets.

Dataset	graphs	label 1 graphs	mean #nodes	mean #edges
NCI1	4110	2057	29.9	32.3
NCI109	4127	2079	29.7	32.1

In this chapter, we illustrate the use of a graph embedding tool to project each graph in the respective datasets into vector space. We then use this

---

[1] ls11-www.cs.tu-dortmund.de/staff/morris/graphkerneldatasets

DOI: 10.1201/9781003218869-9

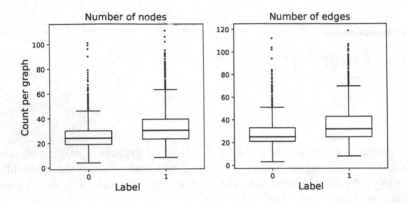

FIGURE 9.2
Node (a) and edge (b) counts per label for the NCI109 dataset.

representation to perform supervised learning, building models from a subset of the compounds, and trying to predict the label of the other compounds. After that, we show how to enrich such representations with features based on the topology of each graph representing the compounds, and we re-visit the supervised learning problem. Finally, we show a few results obtained via unsupervised learning.

9.2 Supervised Learning with Embedded Graphs

In Chapter 6, we saw various techniques to embed nodes of a given graph in a vector space. Here, we consider embeddings for each compound in the NCI1 and, respectively, NCI109 datasets, with the compounds represented as graphs. Thus, we seek vector space representations for these graphs. This can be achieved via methods known as graph kernels (for example, by counting motifs), but those are typically based on handcrafted features that may not generalize well (see Section 6.6 for a short discussion on these methods). Instead, we use an unsupervised, neural embedding framework known as **Graph2Vec**[2] to represent the compounds in a vector space. It is based on the ideas used in **NLP** that we already discussed in Section 6.3 in the context of node embedding algorithms such as **Node2Vec**. Here, the entire graph is the analog of a "document" and rooted subgraphs around each node in the graph, including the node labels, are analogous to the "words" making up the document. Note that this method can be applied to graphs of arbitrary size.

---

[2]github.com/benedekrozemberczki/graph2vec

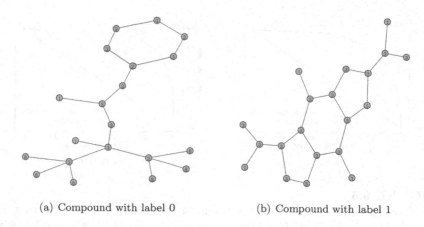

(a) Compound with label 0   (b) Compound with label 1

FIGURE 9.3
Two compounds from the NCI109 dataset.

For the problem of supervised learning using graph embeddings as feature vectors, we experimented with three different representations: (i) high dimensional (1024), (ii) high dimensional reduced to lower dimensional (64) via **UMAP**, and (iii) low dimensional (64). For both datasets, results were better with scenario (ii) as summarized in Table 9.4 where we report the accuracy when building random forests using 80% of the data, and testing on the remaining 20%.

TABLE 9.4
Accuracy over test set for random forest models using various approaches to build feature sets from graph embeddings. A random model has expected accuracy around 0.5 since both datasets are almost perfectly balanced.

Dataset	1024-dim embedding	1024 to 64-dim	64-dim embedding
NCI1	0.623	0.798	0.619
NCI109	0.669	0.837	0.645

We use the **UMAP**-based method to build feature vectors and we show the ROC curves for both datasets in Figure 9.5, again using 80% of the data for training random forest models. In both cases, the results are much better than random, with **AUC** (the area under the **ROC** curve) values around 0.9. We also report the 90% confidence intervals obtained via bootstrap sampling. (Recall that **AUC** can be interpreted as the probability that the model ranks a randomly chosen positive example higher than a randomly chosen negative one.)

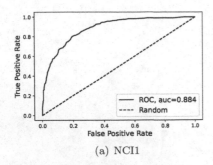
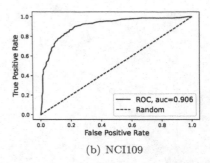

(a) NCI1   (b) NCI109

FIGURE 9.5
ROC curves for the test data set using 64 dimensional feature vectors obtained via embedding in 1024 dimensional space and reducing dimension with **UMAP**. Using bootstrap re-sampling, we obtained a 90% confidence interval of [0.864, 0.903] for the **AUC** with (a) NCI1 and, respectively, [0.889, 0.923] with (b) NCI109.

Adding More Graph-based Features

For each dataset, we can define graph-based features for each compound. This includes values such as the number of nodes, number of edges, edge density, node degree distribution, transitivity, assortativity and more. Details of the computations are given in the accompanying notebook. For the two datasets, we added up to 56 such features to the 64-dimensional representation we have for each compound, and we re-ran the supervised learning experiment, building random forests using 80% of the data.

The corresponding **ROC** curves we obtained are shown in Figure 9.6, showing slightly better results than the ones we obtained using the embedding-based features alone; however, we see some overlap in the 90% confidence intervals for the **AUC**. When using random forests, one can obtain measures of feature importance from the models. In Figure 9.7, we show some of the top features for the NCI1 dataset, plotting their respective distributions over compounds with labels 0 and 1 respectively. From the left plot, we see that label 1 compounds tend to have slightly more degree-3 nodes than the ones with label 0. On the right plot, we show an embedding-based feature. Those distribution are often very close, but we saw that a combination of several such features can bring some benefits.

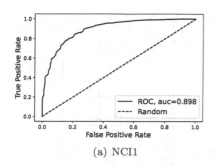
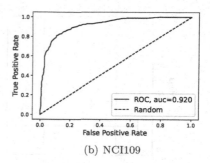

(a) NCI1   (b) NCI109

FIGURE 9.6
ROC curves for the test data set using 64 dimensional feature vectors obtained via embedding in 1024 dimensional space and using **UMAP**, as well as 56 graph-based features. Using bootstrap re-sampling, we obtained a 90% confidence interval of $[0.879, 0.915]$ for the **AUC** with (a) NCI1 and, respectively, $[0.905, 0.936]$ for (b) NCI109. Those are slightly higher but intersect with the ones obtained using only the embedding features.

9.3 Unsupervised Learning

Embedding graphs can also be used to find similar graphs via clustering in the embedded space. For the NCI1 dataset, we took the 64-dimensional representation that gave us good results for supervised learning, and we applied a simple k-means clustering algorithm with $k = 10$ as the number of clusters. We summarize the results in Table 9.8, where we order the clusters in terms of the proportion of label 1 compounds. From those results we see that we obtained a few small clusters with a large proportion of label 1 compounds, and larger clusters with mixed populations. In the accompanying notebook, we further illustrate this via a 2-dimensional colour projection. Those results are very stable when we re-run the k-**means** clustering algorithm several times. We also repeated the same process with $k = 12$ and $k = 15$, and we obtained similar results: a few small clusters containing mostly label 1 compounds. In Table 9.9, we show some graph-based features for the smallest cluster containing 10 compounds, all with label 1. We see that those graphs are very similar in terms of the number of nodes, edges, nodes of degree 1, 2 or 3 and assortativity.

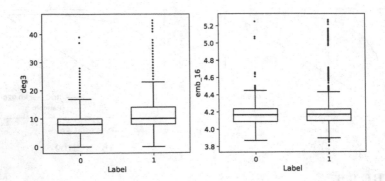

FIGURE 9.7
Showing some top features for the NCI1 dataset model. We compare their distributions over compounds with different labels. The feature in the left plot (a) is graph-based, and the features in the right plot (b) is embedding-based. In both cases, we see a slight difference in the distributions with respect to the two compounds.

9.4   Recommended Supplementary Reading

The dataset's website and source references for NCI1 and NCI109 are:

- K. Kersting, N.M. Kriege, C. Morris, P. Mutzel, M. Neumann, "Benchmark Data Sets for Graph Kernels", 2016.
  `ls11-www.cs.tu-dortmund.de/staff/morris/graphkerneldatasets`

- N. Wale, G. Karypis. "Comparison of descriptor spaces for chemical compound retrieval and classification", *Proc. of ICDM*, pp. 678–689, Hong Kong, 2006.

The reference for **Graph2Vec** is:

- A. Narayanan, M. Chandramohan, R. Venkatesan, L. Chen, Y. Liu, "graph2vec: Learning distributed representations of graphs", *MLG 2017, 13th International Workshop on Mining and Learning with Graphs.* (Graph2Vec).

TABLE 9.8
Clusters in the NCI1 dataset found via *k*-**means** with $k = 10$. We order the clusters with respect to the proportion of compounds with label 1.

cluster size	proportion with label 1
41	1.000
10	1.000
47	0.979
46	0.935
18	0.833
365	0.800
594	0.574
1,380	0.458
1,453	0.398
156	0.378

TABLE 9.9
Some characteristics of the compounds in the smallest cluster found with *k*-**means** in the NCI1 dataset, all of which have label 1. Label 1 compounds tend to have more nodes and edges, and this cluster is a good illustration of this fact.

#nodes	#edges	assortativity	#deg1	#deg2	#deg3
82	97	−0.465995	10	36	32
88	103	−0.494124	12	38	34
78	93	−0.513733	8	35	32
82	96	−0.467626	11	35	33
86	101	−0.535722	11	37	35
78	93	−0.513733	8	35	32
80	94	−0.494897	10	35	32
79	93	−0.506528	10	34	32
80	95	−0.506849	9	35	33
90	105	−0.477663	13	39	33

10

Network Robustness

In earlier chapters, we discussed various measures describing graphs such as centrality measures of their nodes, community structure, degree distribution, or degree correlations. A natural followup question is how these structural properties can help us to better understand the behaviour of the system modelled by these networks and the processes run on them.

One of the important dynamic properties of systems described by networks is their **robustness**, which is also sometimes referred to as **resilience**. This property is related to a rather vague question of what fraction of nodes in the network are closely connected. There are several ways to formalize this notion. In this chapter we will use a measure called the **order parameter**, which is defined as the fraction of nodes of the network that belong to its largest component. In order to test how robust the network is, we will remove some set of nodes before investigating the size of the largest component. Hence, there are two natural ways to report the fraction of nodes that belong to it: as a fraction of nodes in the original graph or a fraction of nodes of the graph after the operation of removing nodes. In the experiments we perform in this chapter, we will use the former conception and always refer to the order of the initial graph *before* any deletions. This definition is less common but allows for cleaner and more easily interpretable visualizations.

Most empirical networks have their **order parameter** close to 1, that is, a graph is either connected or almost all nodes belong to the giant component. We will call a network robust if this property still holds if some of the nodes from the network are removed. Indeed, it is often a desired property of the network to be resilient to either random failures or deterministic attacks that can be viewed as splitting the graph into several smaller disconnected networks when some of its nodes are removed.

Let us consider some practical examples of networks in which resilience is obviously a desired property.

1. Computer network: two nodes (computers) in this network should stay connected even if some of the nodes in the network are removed (for example, some servers go down due to a hardware malfunction).

2. Road network: travellers should be able to travel between any two locations (nodes) even if some other locations are removed from the network (for example, due to road-works).

DOI: 10.1201/9781003218869-10

3. Power grid network: it is crucial for the network to stay fully con-
    nected and operational so that it is possible to transfer energy from
    any source (that is, a power plant) to any destination (for example,
    a factory), even if some random failures of power stations occur.

When analyzing network robustness, there are two typical scenarios that
may be considered.

1. Random removal of nodes: in this case we assume that some force of
    nature or natural process affect the network (for example, a server
    in a computer network goes down due to some hardware error or
    due to its age).

2. Targeted node removal: in this case we assume that some adversary
    selects nodes from the network with the goal of disconnecting it as
    fast as possible (for example, hackers may target several key servers
    in a computer network with a denial of service attack).

In this context, we will consider the following natural questions related to
network robustness.

1. Given a network and its characteristics, how prone is it to random
    or targeted node removal?

2. How much does the speed of deterioration of network's order pa-
    rameter differ between random and targeted node removal?

3. What characteristics of nodes make them suitable for a targeted
    attack? In particular, it is natural to consider removing nodes with
    high degree or with high betweenness centrality, but the question is
    if some global characteristic of the network influences the choice of
    which one should be used.

We start our experiments with illustrating these concepts on an empirical
network representing the power grid graph on the Iberian peninsula (Section 10.1). In order to better understand the interplay between network robustness and some global structural parameters, such as the power-law exponent, assortativity, and the presence of community structure, we then analyze
selected artificially generated networks (Section 10.2).

10.1   Power Grid Network on the Iberian Peninsula

We start by considering the electric grid network from the Iberian peninsula
that we already worked with in Chapter 1. We present the plot of this network
in Figure 10.1. The corresponding undirected graph is connected, has $n = 1,537$ nodes, and the mean node degree equal to $\langle k \rangle = 2m/n \approx 2.553$. This is

a typical example of a spatially embedded graph in which nodes are adjacent only if they are close to each other in the geographical space. As a result, nodes of large degree are rarely present in such networks and they typically do *not* exhibit power-law degree distribution (see a longer discussion about such distributions in Section 2.4). Our first task is to check how robust this network is.

FIGURE 10.1
Giant component of the Iberian peninsula power grid graph.

As discussed in the introduction, we consider the **order parameter**, that is, the fraction of nodes in the giant component of the original graph, as a function of the number of nodes removed under the following three scenarios: random removal of nodes, removal of nodes with the largest degree, and removal of nodes with the largest betweenness centrality. The second and the third scenarios use a greedy approach assuming that after removing a node, the corresponding parameter (the maximum degree or the maximum betweenness) is recalculated for the remaining graph. In case of a tie, we randomly select one of the nodes with the largest parameter to remove.

In Figure 10.2 we show the results of a simulation where we repeatedly undertake the three node removal, until all the nodes are removed from the network. As expected, observe that using a targeted attack makes the order parameter drop much faster than the random removal of nodes. Indeed, in geometric graphs, such as the one we study, we do not have long links. This means that one may try to remove a group of nodes in a specific geographic region and disconnect a large part of the graph from the rest by removing a relatively small set of nodes. That is why adversarial attacks are so effective for such networks. One may also get lucky and reproduce this strategy by random removal but it is highly unlikely.

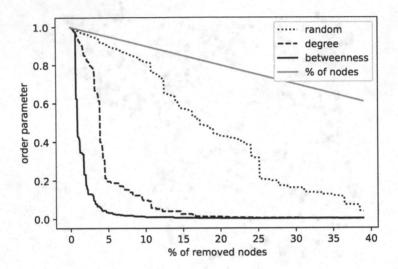

FIGURE 10.2
The **order parameter** as a function of the fraction of nodes removed from the power grid of Iberian peninsula graph. The straight grey line indicates the fraction of nodes remaining in the graph (which is a trivial upper bound for the **order parameter**). The maximum fraction of removed nodes considered is 40%.

In the accompanying notebook we additionally analyze the distribution of cluster sizes, assuming that 5% of nodes are removed from the network under the maximum degree and the maximum betweenness attacks. It is interesting to note that the maximum betweenness attack does not leave large components in the resulting network. On the other hand, the maximum degree attack produces many more small clusters but leaves several relatively large clusters that are kept connected by relatively low-degree, but high-betweenness nodes.

10.2 Synthetic Networks

Let us now move to the analysis of artificial graphs.This will allow us to better understand how global characteristics of networks influence their robustness. We will consecutively consider the following models: **Watts–Strogatz** model, **Chung-Lu** model with the power law degree distribution, assortative/disassortative networks exhibiting degree correlations, and **ABCD** networks with community structure. In all experiments, we consider networks with 2,000 nodes and approximately 4,000 edges.

Watts–Strogatz Graph

In this section we consider the **Watts–Strogatz** model defined on the one-dimensional lattice ring with rewiring parameter p ranging from 0 to 1. This model produces a synthetic network with a small average path length and a large clustering coefficient (see Section 1.11 for a definition and discussion about this graph parameter). For a given number of nodes n and desired average degree $\langle k \rangle$ (which is assumed to be an even integer), we generate the **Watts–Strogatz** graph as follows. We label n nodes with non-negative integers, that is, labels are from the set $L := \{0, 1, \ldots, n-1\}$. We start with a regular ring lattice on n nodes, that is, for each node $i \in L$, we put an edge between node i and node

$$j \in \{i-1, i-2, \ldots, i - \langle k \rangle/2\} \cup \{i+1, i+2, \ldots, i + \langle k \rangle/2\}$$

(using modular arithmetic, mod n) such that node i has degree precisely $\langle k \rangle$. For each edge ij, $i < j \leq i + \langle k \rangle/2$ (as before, mod n), we call i the *leader* and j the *follower* of that edge. Then, we revisit all edges again and for each edge ij we independently rewire the follower node of that edge with probability p, that is, we remove edge ij and replace it with edge $i\ell$, where ℓ is chosen uniformly at random from all nodes. Since we aim for a simple graph, we avoid creating loops and parallel edges by resampling node ℓ, if needed. The parameter p allows us to interpolate between the regular lattice (for $p = 0$) and a random graph that is similar to the binomial random graph $\mathcal{G}(n, q)$ with $q = \langle k \rangle/(n-1)$ (for $p = 1$). The lattice has a large average path length and a large clustering coefficient whereas the random graph has a small average path length and a small clustering coefficient. The "sweet spot" is somewhere between these two extremes, a graph that already has a small average path length but which still has a relatively large clustering coefficient.

As one can observe from Figure 10.3, the parameter p has a significant influence on random node removal. The more long-edges are present in the graph the more difficult it becomes to split the graph into small connected components. On the other hand, for this graph the maximum betweenness attack is relatively efficient but its effectiveness decreases with p. For the maximum degree attack we observe that parameter p has a less significant and non-monotonic effect. The reason for this is that in the **Watts–Strogatz** graph nodes have similar degrees. For instance, it is interesting that the original ring ($p = 0$) is more robust than the random graph for the maximum degree attack. Can this be explained? Yes—in order to decrease the order parameter, one needs to remove consecutive nodes of the lattice. But with the maximum degree attack, once a node is removed, neighbouring nodes will decrease their degrees and, as a result, will not be selected by the greedy algorithm—it will not remove them too soon. The algorithm will remove nodes in some other parts of the graph instead of trying to make a "hole" in the ring.

Power Law Graphs

In this subsection we consider **Chung-Lu** random graphs which generate graphs with a given expected degree sequence following power-law. (See Section 2.5 for details of this model.) We experiment with various power law exponents in the range from 2 to 4. In Figure 10.4, we show how the **order parameter** behaves for different power law exponents. We independently investigate the three scenarios of interest: random, the maximum degree, and the maximum betweenness attacks. We observe that the smaller the exponent's degree is the faster the **order parameter** of the graph decreases in all scenarios. Note that when the exponent is equal to 2 we initially have a significant fraction of isolated nodes in the graph (as it is generated using the **Chung-Lu** random graph which only preserves the expected degree process). Let us also note that the maximum degree and the maximum betweenness attacks have similar efficiency as the considered graph is random in which edges occur independently and so large degree nodes also have large betweenness. However, as in the previous examples, the maximum betweenness attacks are slightly more efficient.

Graphs with Degree-degree Correlations

In this subsection we take the graph from the previous subsection with the power exponent equal to 3 and perturb it using the **Xulvi-Brunet and Sokolov** algorithm to increase and, respectively, decrease its assortativity with $q = 3/4$. (See Section 4.6 for details about this algorithm.) In Figure 10.5 we show the result performed on the graphs from the same experiment that we performed in the previous subsection. One can observe that with random attacks, the disassortative network is most robust. The opposite effect is observed for the maximum degree and the maximum betweenness attacks for which assortative networks seem to be the most robust. The reason for this is that in both scenarios we quickly destroy the *backbone* that consists of high-degree and high-betweenness nodes. However, even if the *backbone* is removed from assortative graph, then low-degree nodes still induce a dense graph as they originally were mutually connected at a higher rate in comparison to disassorative graph. The same justification explains the reason why the maximum betweenness attacks are slightly more efficient than the maximum degree attacks.

Graphs with Community Structure

In the last experiment, we consider the **ABCD** graphs with the mixing parameter $\xi \in \{0.0001, 0.01, 0.1, 1.0\}$ that controls the level of noise. The details of how this model generates synthetic graphs with community structure are provided in Section 5.3. The specific details of the graph generation process used for this specific experiment can be found in the accompanying material from the book's web site.

The results of experiments are presented in Figure 10.6. The general conclusion is that the maximum degree and the maximum betweenness attacks are more efficient for these graphs than for graphs that do not exhibit community structures that we considered in earlier sections. However, the exact value of ξ does not have a very significant influence on the effectiveness of either the maximum degree or the maximum betweenness attacks, provided that it is large enough (specifically, $\xi = 0.1$ and $\xi = 1.0$ yield very similar results). The significant differences can be observed for very low values of the mixing parameter ξ (pertaining to very strong communities).

As in the earlier examples, the maximum betweenness attack is more efficient than the maximum degree attack. It is especially visible for $\xi = 0.01$ where the maximum betweenness attack very quickly eliminates nodes that connect communities (despite the fact that they might not have large degrees).

10.3 Conclusion

Let us summarize what we have learnt in this chapter:

1. Clearly, there might be some artificially generated networks that do not have this property but, as a general rule, using the maximum degree or the maximum betweenness attack is much more efficient than random node removal in making the **ordered parameter** of the resulting network to drop.

2. If the network exhibits power law degree distribution, then the lower the power law exponent the more effective attacks are.

3. If the network exhibits high assortativity, then it is more robust to the maximum degree and the maximum betweenness attacks than highly disassortative networks.

4. In general, the maximum betweenness attacks are expected to be more efficient than the maximum degree attacks.

Finally, let us note that we have only used a few simple centrality measures to select nodes for our attacks (the maximum degree and the maximum betweenness, respectively). In practice, it is possible to perform more sophisticated optimization process ("looking ahead" instead of applying a simple greedy algorithm) for a given network that selects nodes that are most influential, given the predefined number of nodes we aim to be removed. Finding such a set of nodes is a complex combinatorial optimization problem and requires techniques and algorithms that are outside of the scope of this book.

10.4   Recommended Supplementary Reading

The Watts–Strogatz model was introduced in the following paper:

- D.J. Watts, S.H. Strogatz. "Collective dynamics of small-worldnetworks", *Nature* 393.6684 (1998): 440–442. (Watts–Strogatz)

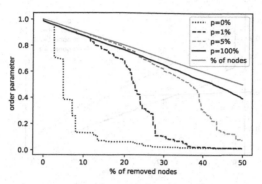

Random node removal

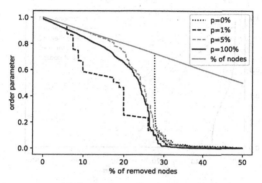

Maximum degree attack

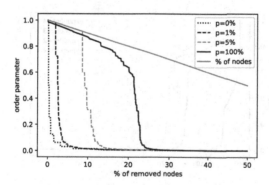

Maximum betweenness attack

FIGURE 10.3
The **order parameter** as a function of the fraction of nodes removed from the **Watts–Strogatz** graph for different values of p. The straight grey line indicates the fraction of nodes remaining in the graph (which is a trivial upper bound for the **order parameter**). The maximum fraction of removed nodes considered is 50%.

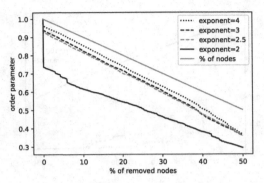

Random node removal

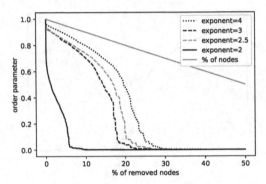

Maximum degree attack

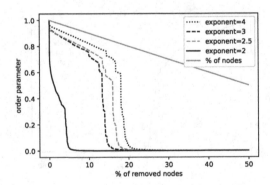

Maximum betweenness attack

FIGURE 10.4
The **order parameter** as a function of the fraction of nodes removed from
the power law graph for different values of power law exponent. The straight
grey line indicates the fraction of nodes remaining in the graph (which is a
trivial upper bound for the **order parameter**). The maximum fraction of
removed nodes considered is 50%.

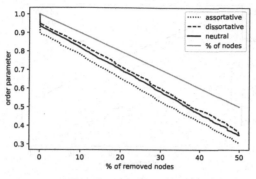

Random node removal

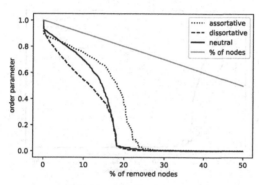

Maximum degree attack

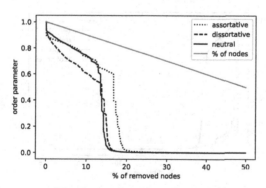

Maximum betweenness attack

FIGURE 10.5

The **order parameter** as a function of the fraction of nodes removed from assortative/disassortative graphs. The straight grey line indicates the fraction of nodes remaining in the graph (which is a trivial upper bound for **order parameter**). The maximum fraction of removed nodes considered is 50%.

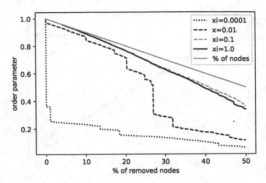

Random node removal

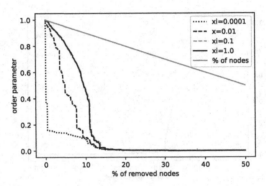

Maximum degree attack

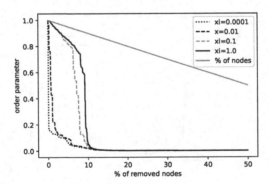

Maximum betweenness attack

FIGURE 10.6

The **order parameter** as a function of the fraction of nodes removed from **ABCD** graphs with varying value of ξ. The straight grey line indicates the fraction of nodes remaining in the graph (which is a trivial upper bound for the **order parameter**). The maximum fraction of removed nodes considered is 50%.

11

Road Networks

In this chapter, we present tools and techniques that can be used to analyze graphs that are embedded in a geographical space. This class of applications is encountered quite frequently in practice. In our example, we concentrate on analyzing the graph representing some network of roads. For the purposes of this chapter, we employ a small part of such a network extracted for the city of Reno, NV, USA, which was obtained from the OpenStreetMap[1] project and is publicly available.

The main goal is to identify intersections in this road network that are exposed to a large volume of traffic. However, let us stress the fact that this particular application is selected to illustrate the process and one should be able to deal with a much broader set of questions using a similar approach. In this particular example, we want to highlight the fact that very often graphs are only a component of more complex mathematical models of some real world phenomena. Such models are frequently quite involved and, as a result, impossible to solve analytically. In such situations a typical approach to analyze them is by using a simulation. In the scenario considered in this chapter, the models that are often used are so-called *agent-based simulations*, where we consider agents (in our case cars) performing some actions (in our case traveling from source to destination location). A comprehensive discussion of how such models are specified and implemented is outside the scope of this book, however, the example we have chosen can be solved with tools and techniques that we have covered in earlier chapters.

11.1 Representing a Road Network as a Graph

Road networks are typically represented as graphs in the following way. Nodes in such graphs correspond to intersections and edges correspond to road segments between the intersections. In the additional material from the book's website, we include code that shows how one can easily extract the necessary data from the OpenStreetMap[2] project to obtain:

---

[1] www.openstreetmap.org/

[2] We used the following excellent package: github.com/pszufe/OpenStreetMapX.jl

DOI: 10.1201/9781003218869-11

1. geographical locations (latitude and longitude) of nodes;
2. weights (road lengths) of edges, along with the road class (determining the traveling speed).

In Figure 11.1 we show an example of a graph representing a road network. However, if a graph is actually embedded in a geographical space we may use this fact and try to do a better job. In Figure 11.2 we illustrate how one can present the graph representing the road network on top of the display of the map. This approach uncovers more details about the structure of the graph. For example, we see that there is an airport at the bottom of the map (and thus there are no roads there) or that there is a river, slicing the map horizontally, with only a few bridges (edges) crossing it. Finally, observe that there is a highway that is represented in pink in the accompanying Jupyter notebook. In the accompanying notebook one may additionally see a picture showing that roads on a highway have the highest speed of travel, as one would expect. It is possible to overlay a graph on top of the map like this using the `folium`[3] package available in Python. In particular, let us mention that in the accompanying notebook one may interact with the plot by zooming in or out, or by moving it.

The graph representing a road network that we are presently considering is directed and strongly connected. It consists of $n = 1{,}799$ nodes and $m = 3{,}963$ edges. In Table 11.3, we present node in- and out- degree distributions. We can see that frequencies of in- and out- degrees for a given value are very similar, which suggests that the majority of the roads are two-way. Also, interestingly, frequencies of in-degree 1, 2, and 3 are similar to each other whereas 4 occurs much less frequently (the same applies for out-degree). In the accompanying notebook, plots of the graph with nodes coloured by their in- and, respectively, out- degree are provided. From this plot one can learn, for instance, that most nodes lying on a highway have in- and out- degree equal to one. This is due to the fact that highways have a number of entries/exits and U-turn points. As a result, the representation of a highway in OpenStreetMaps project has many nodes having degree one to properly represent its shape.

Similarly, in Table 11.4 we list fractions of edges with different driving speeds. We see that most of the roads have low speed limits (they are most likely located in some dense residential zones), and the fastest roads constitute less than 6% of edges. As observed above they form a highway.

11.2 Identifying Busy Intersections

The objective of our analysis is to identify the intersections in the road network that are expected to experience a lot of traffic. In order to perform this analysis

---

[3]`python-visualization.github.io/folium/`

FIGURE 11.1
Graph representing the system of roads of Reno, NV, USA.

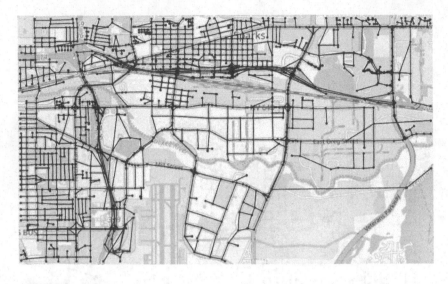

FIGURE 11.2
Road graph plotted on top of the map of Reno using the `folium` package.

TABLE 11.3
Distribution of node in- and out- degrees for the graph representing road
network in Reno, NV, USA.

degree	in- %	out- %
1	32.07	31.80
2	25.79	26.13
3	30.79	30.96
4	11.34	11.12

TABLE 11.4
Distribution of driving speed limits for the graph representing road network
in Reno, NV, USA.

speed	%
40	59.73
50	13.21
70	18.93
90	2.46
120	5.67

we assume that citizens want to travel between any two randomly picked intersections in the considered road network via a shortest path linking them (note that there may be more than one path with the shortest length between any two intersections). This is, of course, a simplified model but it should still be able to predict the possible outcome. A more realistic model would require additional information about the behaviour of people living in the city such as their home location, their workplace, number of cars, etc. If such external knowledge is provided in a convenient format, adjusting the model should be straightforward.

Our goal is to show how sensitive the results of the analysis are with respect to the level of detail reflected by the graph. Therefore, we present three scenarios with increasing levels of realism:

1. assuming that each edge in the road graph has the same length and travel time;

2. taking into account real road lengths but assuming that the driving speed on each edge is the same;

3. taking into account real road lengths and differentiating driving speeds between roads (in particular, one may drive faster on a highway than on a gravel road).

Of course, even the third scenario that we consider is still lacking many real-life aspects that are potentially important in practice and might affect the result significantly. Here are some natural extensions we ignore in the analysis:

1. non-uniform distribution of the source and the destination locations of travel (in particular, traffic generated from outside of the area we have selected for analysis is completely ignored);

2. the number of lanes on each road;

3. relationship between traffic on a road and effective average driving speed;

4. road usage restrictions for certain classes of vehicles;

5. effect of street lights;

6. restrictions on turning on intersections.

We left out these details from the analysis to keep the example simple enough. However, it is certainly possible to extend the analysis with these details, which could constitute an interesting computational project.

Let us now discuss how the analysis we want to perform can actually be executed. The first idea that comes to mind is to write a simulator of cars driving around the city. This approach is certainly feasible (and, in fact, two of the authors of this book have several research papers applying such a methodology), however, it is quite expensive to implement and execute. Fortunately, the question we want to answer can be quite well approximated by *betweenness centrality*, which we introduced in Section 3.3. The reason behind it is that, since we assumed that the source and the destination nodes are sampled uniformly at random from the set of all nodes and that people select a shortest path to get to their destinations, the betweenness centrality directly translates to the probability a given node will be visited on such a route.

In the accompanying notebook we perform this analysis for the three notions of node distance, as discussed above: ignoring road lengths, taking into account road lengths, and taking into account both road lengths and speed limits. Before we discuss the result, let us note that under all three distance measures most of the nodes have very small betweenness and only a few nodes have large betweenness. (This can be investigated on the betweenness distribution histograms presented in the notebook.) This means that, indeed, one may expect that certain intersections in the city create a "bottleneck" in the road system and could become very busy.

In Figure 11.5 we present the results of the analysis when road lengths are ignored. In this figure and in the following figures, large black circles represent the top 1% of the busiest intersections, and small black circles indicate the top 10% of the busiest intersections. One non-surprising conclusion is that the busiest intersections are located in the center of the map; in particular, around the bridges over the river. However, we can also spot some deficiencies. For instance, some local roads located at the top of the map seem to be indicated as relatively busy, which does not sound realistic. Therefore, in Figure 11.6 we incorporate information about road lengths in the analysis. Now, we see that the points are less scattered and lie along the main roads. We also notice that the nodes in the very center of the map in Figure 11.5 are now slightly less busy as they are quite distant from each other and so are not used as much as would seem when ignoring the road lengths. Still, it seems that this analysis could be improved as we note that, somewhat surprisingly, highways are not utilized as much as one would expect. This is "fixed" in Figure 11.7 where we additionally take into account the distance on the graph using travel time (that is, the road lengths divided by road speeds). This change, indeed, leads to a solution where the busiest nodes lie along the highway, as they should.

In this experiment, we observe that relatively small changes to the setting of the problem might lead to significantly different conclusions. Such an iterative process of refining the analysis with additional assumptions is quite typical in the daily work of a data scientist. The most challenging part is to decide, following the principle of *Occam's razor*, when to stop adding them. Fortunately, as we have tried to show in this chapter, experimenting with graph mining techniques is really easy, especially when you have a helpful visualization at hand, and we hope you enjoyed it as much as we did!

FIGURE 11.5
The busiest intersections assuming all road sections have the same weight in the road graph of Reno.

FIGURE 11.6
The busiest intersections taking into account road lengths in the road graph of Reno.

FIGURE 11.7
The busiest intersections taking into account travel time in the road graph of
Reno.

11.3   Recommended Supplementary Reading

An introduction to simulation of transportation systems can be found in the
following textbook:

- D.P.F. Mller. "Introduction to Transportation Analysis, Modeling and Simulation", Springer, 2014.

Index

Printed in the United States
by Baker & Taylor Publisher Services

Printed in the United States
by Baker & Taylor Publisher Services